10 GREATEST SHIPS
OF THE
ROYAL NAVY

JOHN BALLARD

AMBERLEY

First published 2015

Amberley Publishing
The Hill, Stroud
Gloucestershire, GL5 4EP

www.amberley-books.com

British Library Cataloguing in Publication Data.
A catalogue record for this book is available from the British Library.

ISBN 978 1 4456 4652 7 (print)
ISBN 978 1 4456 4653 4 (ebook)

Typeset in 10pt on 12pt Sabon.
Typesetting and Origination by Amberley Publishing.
Printed in the UK.

Contents

Introduction

For more than 150 years it was the world's most powerful force, a mighty presence in every corner of the globe.

From Newfoundland to New South Wales and Cyprus to Ceylon, the Royal Navy was there: protecting British interests, projecting British power and maintaining British prestige.

Britain's rise to global pre-eminence owed as much to the navy as it did the Industrial Revolution. In fact, the two were inextricably linked.

The goods mass-produced in British factories could be shipped across the world unmolested as, for a century and a half, that cliché about Britannia ruling the waves was true. Controlling the international shipping lanes meant Britain controlled world trade and could make global affairs dance to a British tune.

Strategic maritime locations – Gibraltar, Malta, the Suez Canal, Aden, the African cape and Singapore – were all in British hands.

Between victory at Trafalgar in 1805 and the withdrawal from 'east of Suez' in the 1960s, the ships of the Royal Navy were ubiquitous.

It was even laid down in an Act of Parliament (Naval Defence Act 1889) that the Royal Navy had to maintain a number of battleships at least equal to the combined strength of the next two largest navies in the world.

It wasn't just the numerical warship advantage that Britain held. Shipyards up and down the land were sending the most technologically advanced vessels of the age down their slipways. Ships that could outrun, outmanoeuvre and out-gun anything their rival nations put to sea. Had it not been for the navy, Britain would most likely have remained insignificant on the global stage instead of the superpower she was to become.

Ever since defeating the Spanish Armada, naval power had been key to keeping would-be invaders at bay, but it was after Trafalgar that Britain became undisputed master of the seas.

In the nineteenth and early twentieth centuries, Britain was a country obsessed with its navy in a way that seems alien today. Images of the navy were everywhere, from Jack Tar caricatures on cigarette cards to school textbooks extolling the virtues of British sea power.

Reams of column inches were given to naval developments and the seemingly never-ending launch of new ships at yards across the United Kingdom. Harland & Wolff of Belfast, Vickers from Barrow, John Brown's of Clydebank, Fairfield of Govan, Cammell Laird at Birkenhead and Swan Hunter in Wallsend were among the household names creating these technological marvels that captured the public imagination.

Gunboat diplomacy is now just a figure of speech, but in Victorian and Edwardian times it was the prime weapon in Britain's armoury that could cow enemies into submission and kept a quarter of the world under the Union flag.

But while the first years of the twentieth century seemed to offer everlasting promise and prosperity for Britain and her navy, rough seas lay ahead.

Germany had been gradually building an industrial output, economy and navy to rival the British for several decades, and though the launch of HMS *Dreadnought* in 1906 rendered all other warships immediately obsolete, the Kriegsmarine quickly closed the gap in the naval arms race. The German fleet might not have been the equal of the Royal Navy but it was now playing in the same league and posed a major threat at the outbreak of the First World War.

For more than a century generations had grown up on stories of the Trafalgar victory and the invincibility of Britannia's fleet, so while the rival armies might become bogged down in the mud of Flanders, the carnage of Ypres and bloodshed of the Somme, Britain expected the knockout war-winning blow would be delivered at sea.

To counter the German naval threat and set up base in a strategic location for operations against the enemy, the British Grand Fleet moved to Scapa Flow in the Orkney Islands at the start of hostilities in 1914.

One of the world's largest natural harbours and occupying a prime position at the entrance to the North Sea, Scapa Flow became a temporary wartime home to the largest collection of maritime might ever assembled.

But although the Admiralty had this awesome arsenal at its disposal, the War Cabinet and naval high command preferred the more risk averse option of staying at anchor and battling the Germans only if the kaiser's ships went on the offensive.

For the first two years of war the adversaries were content to watch each other from afar and it wasn't until May 1916 that they engaged.

To lure the British from the safety of Scapa Flow, the Kriegsmarine sent five battlecruisers out of port into the North Sea – and the Battle of Jutland, off the coast of Denmark, ensued.

While both sides inflicted damage on each other, the result was effectively a score draw – with the Royal Navy winning on away goals.

The British lost more ships, but with a larger fleet it was a hit they could arguably afford to absorb, and this made the Germans realise they simply didn't have the numbers to deliver a decisive blow. They stayed in port for the rest of the war and eventually succumbed to a combination of a Royal Navy blockade that cut food supplies and the wearing down of their armies on the Western Front.

While victorious, the First World War was ultimately bittersweet for Britain and the navy. The fleet remained the largest in the world but the war left Britain's people and finances exhausted, and limits imposed by the League of Nations on the number of warships the great powers were allowed to build levelled the playing field so that the United States, France and Japan closed the gap on British sea supremacy.

When war engulfed Europe in 1939, for the second time in a little over twenty years, the Royal Navy again moved the Home Fleet to Scapa Flow.

This time round, though, there was no sizing up adversaries from afar. The conflict engaged the Senior Service right from the off and it saw constant action across the world over the next six years. Despite numerous setbacks and heavy losses, the king's ships played a crucial role in the Allied victory and, although now second in size to the US Navy, the Royal Navy had stood firm and more than done its duty.

With the Nazi and Japanese threat vanquished and the colonies of the British Empire granted independence in the post-war years, the navy had a fight on its hands to stay relevant but it was a fight that it won.

There was no longer a vast empire to protect but, as a key partner of the United States, thwarting the red menace and keeping the Soviet Union at bay became the fleet's new raison d'être.

During the Korean War of the early 1950s, the Royal Navy showed just how well it was suited to its new role, deploying aircraft carriers and a variety of other vessels as part of the United Nations' campaign to combat communism on the Korean peninsula.

Having been at the forefront of seafaring developments for hundreds of years, the navy wasn't about to be left behind in the second half of the twentieth century and, despite dwindling power on the world stage and ever-tightening Treasury belts, Britain kept up technologically, if not numerically, with the United States and Soviet Union. Nuclear submarines, amphibious landing ships and ever more advanced aircraft carriers, destroyers and frigates continued to roll down British slipways and these new vessels formed the cornerstone of NATO's deterrent against Soviet expansionism.

Although still able to pull up a seat at the global top table, Britain's power began to wane even further and more rapidly during the 1960s so that, by the end of the decade, there were no longer any Royal Navy bases east of the Suez Canal – or anywhere outside the British Isles and Gibraltar for that matter – apart from on a group of remote windswept islands in the South Atlantic.

And it was in the waters of these islands, 8,000 miles from the United Kingdom, that the navy would see arguably its finest hour since Trafalgar.

The Falklands Conflict of 1982 was a throwback to an earlier age. A British fleet was dispatched to drive out foreign aggressors and claim – or in this case reclaim – territory for the Crown.

Having been weakened by savage defence cuts just a year earlier, the Senior Service nevertheless assembled its largest armada since the 1956 Suez Crisis.

Despite fighting a determined enemy in a hostile environment with limited air cover, and losing ships and men at regular intervals, the Royal Navy won a magnificent victory, successfully combating the Argentine naval and air threat to land ashore the troops who would defeat Argentina's ground forces.

Britain's triumph in the conflict captured the public imagination, transformed Margaret Thatcher's political fortunes and unleashed a tidal wave of patriotism not seen since the Second World War.

Reinvigorated by its key role in the Falklands, the navy was, temporarily at least, spared further cuts, and vessels that had been destined for sale or the breaker's yard stayed in service.

As the Cold War drew to a close, though, the axe fell and Her Majesty's ships were pruned back to such small numbers that it would've had Edwardian seafarers choking on their rum.

Yet, regardless of the reduction in size, the navy continued to be at the centre of world events, seeing action in the Gulf War, Yugoslavia, Kosovo and the Iraq War, as well as taking part in various humanitarian relief efforts and quietly going about its business of protecting British interests.

Smaller now than it has been for hundreds of years, the present and future are still bright for this unique fighting force.

Two new giant aircraft carriers – the largest ships ever built for the Royal Navy – are under construction in Scotland and due to enter service early in the next decade.

Astute Class nuclear submarines are being built in Barrow, while on the Clyde designs and plans for a new class of frigate to replace the ageing Type 23s are progressing.

There'll also be Trident successor submarines to build and, before too long, new mine hunters and patrol vessels.

So while the media and public gaze is very much on the size of the twenty-first-century navy, to focus on this for too long is to slightly miss the point.

The Royal Navy remains one of the most technologically advanced and respected navies in the world and possesses capabilities and weapons systems most other country's fleets can only dream of.

With new carriers, frigates and submarines on the horizon, these capabilities will increase and Britain will continue to operate one of only a handful of true blue water navies – able to dispatch vessels at short notice to any sea or ocean.

While the future is undoubtedly bright, this book is more concerned with the Royal Navy's glorious past.

Covering a period of some 200 years, it tells the stories of what, for my money at least, are the ten greatest ships to have served under the White Ensign.

Whether through war service, cutting-edge design that revolutionised naval warfare, versatility, or a combination of all three, the vessels featured all left an indelible mark on the Senior Service.

While obviously focussing on maritime affairs, the next ten chapters also weave in British political and social history, for what happened in politics and society were a catalyst for and result of what took place at sea.

CHAPTER ONE

HMS *Victory*

HMS *Victory*
Built: Chatham Dockyard, Kent
Weight: 4,000 ton
Served: 1765–1812, but remains in commission to the present day

She might not have seen active service since 1812, but for the past two centuries HMS *Victory* has remained one of the Royal Navy's most potent weapons.

Her role at Trafalgar transformed *Victory* from a mere warship into a national symbol – an icon that embodied British mastery of the waves and the power and supremacy of Britain's navy.

Generations of school children, and every rating and officer to serve on His or Her Majesty's ships from 1805 onwards, has, at least in some small way, absorbed the spirit of Trafalgar and been taught to channel it as inspiration.

If evidence was ever needed of the almost mystical hold *Victory* and Admiral Horatio Nelson have had on the nation's psyche, it's right there in central London – Nelson stands surveying the capital atop a 150-foot column in the middle of Trafalgar Square.

Victory wrapped a cloak of invincibility around the navy and created an aura just as fearsome and powerful as any gun, missile or aircraft. While the cloak is now much smaller, and has been worn and threadbare at times, it's still there. After all, the Senior Service hasn't lost a war since *Victory* draped it over Britannia's shoulders. Perhaps the greatest tribute to her legacy is that she's still around today, flying the White Ensign as a commissioned ship and visitor attraction at the Royal Navy's home in Portsmouth.

This combined role of museum and ship in commission – flagship of the First Sea Lord no less – is a nod to the role *Victory* continues to play now as well as celebrating her illustrious past.

Before 'the' HMS *Victory*, five earlier ships with the same name had served in the Royal Navy. The first was a former merchant vessel that helped see off the Spanish Armada in 1588 and the most recent a 100-gun first-rate ship of the line launched in 1737, but tragically wrecked off the Channel Islands, with the loss of her entire crew, seven years later. Given such an ignominious end to the last vessel of that name, the Admiralty could have been forgiven for shying away from it.

In the late 1750s hostilities yet again erupted between two old adversaries as Britain and France vied for control of Canada, the American colonies and the Caribbean in what became known as the Seven Years' War. To put Britain on a firmer war footing, Parliament voted to build twelve new ships of the line, including a 100-gun first-rate, with the aim of quite literally blowing the French out of the water.

At the height of the conflict in 1758 – incidentally the year of Nelson's birth – a new dry dock was prepared at Chatham Dockyard in Kent to construct the as yet unnamed centrepiece of Britain's much expanded and modernised fleet.

Designed by Sir Thomas Slade, surveyor of the navy from 1755 to 1771, the plans for *Victory* were approved by the Board of Admiralty in June 1759 and the keel laid on 23 July that year.

Tour a modern day shipyard and you'll see prefabricated module blocks made of welded steel, cut and shaped by advanced computerised machinery – a completely alien world to John Lock, the shipwright in charge of building *Victory*.

Yet to take a Georgian shipbuilder like Lock back 250, 500 or even 1,000 years before his time, to look at ship construction in eras far before his own, would be to show him a relatively familiar scene. For *Victory* was made out of wood and built with hand-tools, a practice little changed in centuries.

Some 6,000 very carefully selected trees, mainly oak felled in the forests of Kent, but with elm for the keel and fir for the upper decks and bulkheads, were used in the construction. Around 250 men set about bringing her to life in what was one of the great pre-industrial shipbuilding achievements.

One year in to the build, *Victory* finally got her name. With the tide of the Seven Years' War turning in Britain's favour following successes on land and at sea, *Victory* seemed an appropriate title for the flagship of a country burgeoning in confidence.

The Royal Navy had pummelled the French fleet at the 1759 Battle of Quiberon Bay, so, as well as selecting a grandiose name to celebrate this triumph and capture the national mood, the actor David Garrick penned the celebrated, sentimental and rousing song 'Heart of Oak' – an ode to the prowess of British seafaring that remains the Senior Service's official march to this very day.

> Heart of Oak are our ships,
> Jolly Tars are our men,
> We always are ready, steady boys, steady!
> We'll fight and we'll conquer again and again.

There was no finer heart of oak anywhere than the one taking shape in the Old Single Dock at Chatham Dockyard yet ironically, especially given the role she was to play four decades later, *Victory* was almost a victim of the navy's success.

With the Seven Years' War all but won, the need for a state-of-the-art first-rate ship of the line was far less pressing, so her construction temporarily ceased. It finally got underway again in the autumn of 1763 and she was eventually launched on 7 May 1765 to much fanfare at a final build cost of £63,176 (around £50 million in today's money).

Then, as now, the launch of a new ship attracted dignitaries, well-known faces, widespread media interest and more than a touch of hyperbole – just look at 2014's naming ceremony for the giant new aircraft carrier HMS *Queen Elizabeth* for proof of how little has changed in that respect.

On *Victory*'s launch day, the *London Public Advertiser* eagerly anticipated the event, telling its readers,

> This day will be launched His Majesty's ship the *Victory*, estimated the largest and finest ship ever built. Several of the lords of the Admiralty, commissioners of the Navy and many persons of quality and distinction are expected to be present, for whose receptions great preparations are making through the town.

Having been built in a dry dock, *Victory* was floated out into the River Medway rather than released in a more dramatic fashion down a slipway but, for the assembled crowd of VIPs, naval top brass and dockyard workers, it was still a wonder to behold.

At 230 feet long, capable of speeds between 8 and 9 knots and weighing in at 4,000 tons, she was one of the most complex moving machines of the time, as advanced for her age as nuclear submarines and aircraft carriers are to ours.

With 100 guns spread out over three main gun decks, capable of firing one and a half tons of iron shot in one devastating broadside, *Victory* was the ultimate expression of maritime menace to Britain's foes.

Many of the ship's statistics made for mindboggling reading. Her thirty-seven separate sails all stretched out would cover an area larger than the size of a football pitch, 26 miles of rope were needed for her rigging and enough wood was used in her construction to clad the Empire State Building one and a half times.

Yet, for all the fearsome firepower and impressive statistics, *Victory* spent the first thirteen years of her life tied up on the River Medway. With no immediate conflict on the horizon, the Admiralty took the decision to place her in reserve – or extended readiness, to use today's naval parlance.

It wasn't until France sided with the colonists in the American War of Independence that she was finally commissioned in February 1778, at an age where most modern day warships would be about halfway through their lifespan.

Upon her commission, *Victory* became flagship to Admiral Augustus Keppel and Britain's Grand Fleet. Having waited well over a decade to see action she got a taste of it just five months in to her active naval career at the first Battle of Ushant.

Twenty-nine British ships of the line, commanded by Keppel, met thirty-two French vessels 100 miles or so west of the island of Ushant, off the coast of Brittany, in July 1778.

What looked like an impending titanic battle between the world's two greatest fleets, however, ended up little more than an indecisive damp squib that was notable only for *Victory* firing her guns in anger for the first time – getting shots away at the French flagship *Bretagne*.

Both sides headed for home having been unable to inflict much damage, although driving wind and heavy rain weren't exactly conducive conditions in which to fight eighteenth-century sea battles.

For the next three years *Victory* served as flagship to three different admirals and won an enviable reputation for her nimble handling and speed. Her pace was further enhanced in 1780 when she was sheathed in copper plating to tackle woodworm and stop the spread of marine growth. The lack of sea life clinging to her hull reduced drag and meant she cut through the water more effectively.

The copper plating was added just in time for *Victory*'s next clash with France – the second Battle of Ushant, in 1781.

A convoy, protected by nineteen ships of the line, sailing from Brest to the West Indies with reinforcements and supplies for French forces in the Caribbean was intercepted by a thirteen-ship-strong British force commanded by Rear Admiral Richard Kempenfelt in *Victory*.

Despite being heavily outnumbered, Kempenfelt's force managed to evade the escort vessels and capture fifteen troop and cargo ships before the French could get their bearings. In many ways, though, it was a missed opportunity: the British didn't have sufficient warships to deliver a decisive blow that could have badly mauled, if not destroyed, much of the French navy and put a stop to Napoleon before he even came to power.

A ship is only ever as good as the crew that serve in it, and this was never truer than in the case of the 800 officers and ratings manning HMS *Victory*.

Just as it was ashore at that time, life in the Georgian navy was tough. Primitive cramped conditions, poor sanitation, strict discipline, dull food, heavy manual labour and repetitive tasks meant it wasn't a role for the faint-hearted.

Many on board weren't there out of choice as the 'impressment' or 'press gang' system was still in operation, whereby men between the ages of eighteen and fifty-five with a seafaring background could, during times of war, be literally dragged off the street or merchant ships and taken into the navy by force.

So many sailors living in such close proximity – 500 crew lived, ate and slept on *Victory*'s lower gun deck alone – meant quarrels and fights were common and diseases such as yellow fever and typhus could quickly spread throughout a ship. The generous rum ration of half a pint per man per day, plus beer or grog, probably contributed to the frequent flair-up of tempers, but it no doubt helped ease the stresses and strains of life at sea too.

Those who did overstep the mark could expect swift, physical punishment – usually in the form of a cat-o'-nine-tails flogging, a penalty carried out in front of the rest of the ship's company in the hope of deterring others from contravening the rules.

However, what may seem, to our twenty-first-century sensitivities, intolerable working and living conditions weren't viewed as such in the eighteenth and early nineteenth centuries. Despite the harsh conditions, a naval career had many benefits over life ashore. The staple navy diet of salt beef and pork, biscuits, oatmeal, cheese and pease might not sound overly appetising but at least serving in His Majesty's ships meant three meals a day, a roof over your head, a steady supply of alcohol and earning a regular wage – more than can be said for many civilians during this period.

Given the pride and trust Britain had in her fleet, being a naval officer or rating was also a role with some prestige and many sailors were quite rightly proud of their seafaring expertise.

Just like today, the gunnery skills and seamanship of the Royal Navy's mariners were second to none and crews were trained until whatever they were tasked with became second nature.

Cannon on *Victory* tended to be manned by a six-man crew – all of whom had specific tasks in this vital role.

The gun captain would aim and fire the weapon, two spikers (one on either side of the gun) were responsible for turning and moving the gun after firing, while the loader would place the cannon and ram the shot and powder home. The sponger had the task of sponging out the gun and clearing it of hot debris and finally the powder monkey – usually the smaller members of the crew, often young boys – collected gunpowder charges from the ship's magazine and ran them to the guns.

Every Royal Navy gun crew was trained to fire and reload their gun, ready to go again, in just 90 seconds – a standard of gunnery unmatched in every other navy of the time, and therefore one of the key factors in establishing British naval dominance.

Compared to modern guns and missiles, cannon seem positively prehistoric but they could wreak destruction, causing horrific deaths and injuries in the process.

Three main types of shot were used in *Victory*'s guns. 'Round shot' was for pounding and penetrating the hulls of enemy vessels, 'dismantling shot' to take out masts and rigging and 'grape shot' was deployed to kill and maim opposing sailors and marines. Staggeringly, the 104 cannons *Victory* mustered at Trafalgar possessed more firepower than all of Wellington's heavy guns at the Battle of Waterloo.

If fired at a ship's weak spot, cannon balls could easily tear a destructive path right through a deck, inflicting numerous deaths and wounds with a single shot.

A twenty-year-old marine, Louis Roteley, who served on board *Victory* at Trafalgar, paints a vivid description of the hell that was battle at sea in this era.

> A man should witness a battle in a three-decker from the middle-deck, for it beggars all description. It bewilders the senses of sight and hearing. There was fire from above, fire from below, besides the fire from the deck I was upon, the guns recoiling with violent reports louder than thunder, the deck heaving and the side straining. I fancied myself in the infernal regions, where every man appeared a devil. Lips might move but orders and hearing were out of the question: everything was done by signs.

Suffering serious wounds in eighteenth- and early nineteenth-century sea battles could be a fate far worse than instant death.

Cannon balls and wooden splinters from blasted hulls often caused severe leg and arm injuries as limbs got caught in the flight paths of projectiles. If you survived the initial impact a more prolonged and painful death could easily await.

Surgery – both on land and sea during *Victory*'s time in service – was rudimentary at best, so even those wounded who survived long enough to get to the operating table didn't have the odds stacked in their favour.

Operations were carried out in the action emergency operating theater set up on the midshipmen's mess tables and sea chests. Because this area was cramped and dimly lit, treatment under these conditions were crude. Using little or no anaesthetic, the surgeon and his mates carried out amputations and removed splinters and musket shot. After amputation the stumps of limbs were sealed with spirits of turpentine. Warm tar was used as a last resort. Surgeons had to be very efficient. In battle conditions they could amputate a limb in one and a half minutes, but anaesthetic and after-care were non-existent. Sadly many men died from shock, loss of blood or awaiting surgery.

Fortunately, in *Victory*'s third skirmish – the 1782 Battle of Cape Spartel – there was little call for surgeons and amputations as she didn't actually fire a shot.

Spain had entered the American War of Independence on the side of the French and American colonists with the aim of capturing Gibraltar – a strategic location at the entrance to the Mediterranean that had been ceded to Britain under the Treaty of Utrecht in 1713.

In an attempt to drive the British out, Franco-Spanish forces put the peninsula under siege by imposing a land and naval blockade, although the naval blockade had been bypassed by the Royal Navy in both 1780 and 1781 and the Rock was resupplied.

In 1782, however, supplies were running short again so the Admiralty dispatched a large convoy of transports escorted by thirty-five ships of the line to deliver the goods. Supplies were landed in Gibraltar successfully and it wasn't until the homeward journey that the Franco-Spanish fleet got close enough to the British to open fire. Admiral Richard Howe, carrying his flag in *Victory*, ordered his ships to fire back but stuck to his orders and returned to England rather than engaging the enemy in a full-on confrontation.

Back home, *Victory* underwent a major refit in March 1783. Her deck armament was modified and firepower beefed up by replacing her 6-pounder guns with 12-pounders, but she didn't take part in any further battles until the next decade.

With Britain and Spain still at loggerheads in the 1790s, and both vying for control of the Mediterranean, further conflict between the two nations was inevitable and it came in 1797 at the Battle of Cape St Vincent – a clash that put Nelson on the path to glory.

Fifteen Royal Navy ships of the line faced a twenty-seven-strong Spanish squadron and, while Admiral John Jervis on board *Victory* was in command of the British operation, Britain's decisive defeat of the Spanish fleet owed much to Nelson, then a commodore on HMS *Captain*.

Nelson gave orders for *Captain* to break away from the line of battle and cut off the enemy's escape route. Other ships followed his example and intercepted the Spanish, killing or wounding hundreds of sailors and capturing a number of vessels including the 112-gun *San Jose*.

Victory herself played an important role in the battle, raking enemy ships with heavy fire and delivering a series of devastating broadsides.

While Nelson was given a knighthood and promoted to Rear Admiral following Cape St Vincent, *Victory* very nearly reached the end of the line, rather than remaining a ship of the line.

Now more than thirty years old and a veteran of four battles, the Admiralty ordered her removal from active service and conversion to a hospital ship to hold wounded prisoners of war before luck or fate, call it what you will, intervened.

In 1799, the 98-gun ship of the line HMS *Impregnable* ran aground off the southern coast of England and couldn't be refloated so, with the Royal Navy now short of a first-rate, it was decided to recondition and retain *Victory* as a warship.

The refit got underway in 1800 but the extent of work required meant that it cost around three times the original estimate. *Victory*'s magazine was lined with copper, the figurehead (the carved decoration on her prow) and masts replaced, large parts of the hull rebuilt, the number of guns increased from 100 to 104 and the black and yellow paint scheme the ship still wears to this day was added.

Ready for service again in 1803, she took part in the ongoing blockade that kept much of the French fleet holed up in Toulon – it was a blockade Napoleon was determined to smash. His theory was simple; destroying, or at least outwitting, the Royal Navy would remove the only obstacle preventing him invading Britain. The French dictator's tactic was to try and lure the British fleet from its home waters so his invasion forces could cross the Channel unmolested and land troops on the south coast of England.

This attempt at enticing the British out of the Channel essentially led to a series of cat and mouse games between the British and Spanish and French navies that came to a head in October 1805 in the course of the action that ensured the names *Victory* and Nelson will live on in British history forever.

Having made a dash from Toulon, the French fleet under Admiral Pierre Villeneuve headed for the West Indies to draw the Royal Navy from the Channel, as per Napoleon's plan. However, the British pursuit foiled this plot and trapped Villeneuve and the Spanish navy in the port of Cadiz in southern Spain. On 18 October, the Franco-Spanish fleet of thirty-three ships of the line broke out from Cadiz and into the inevitable titanic confrontation with the Royal Navy's twenty-seven front-line vessels.

Nelson, flying his flag on board *Victory*, formed his ships into two columns and sailed towards the enemy giving the immortal signal: 'England expects that every man will do his duty.'

Naval orthodoxy at the time was to attack an enemy fleet in one single line of battle but Nelson tore up the rule book and broke the Franco-Spanish forces into three separate formations so as to completely surround one third and force them to fight to the end, rather than giving them the opportunity to flee.

Victory steamed into action, cutting the enemy line and delivering a destructive broadside from almost point-blank range into the stern of Villeneuve's flagship *Bucentaure*.

With battle raging all around, and the British mercilessly pounding every French and Spanish vessel in their sights, Nelson was felled by a French marksman aboard *Redoutable*. The shot hit his shoulder, pierced his lung and lodged in his spine. 'They have done for me at last, Hardy. My backbone is shot through,' said the great man as he was taken below.

By the time he died a little over three hours later the Royal Navy had secured the greatest triumph at sea in history and set Britain on course for global domination. Seventeen enemy vessels were captured and one destroyed without the loss of a single British ship.

Victory, losing fifty-seven men, was badly damaged in the battle and had to be towed to Gibraltar for basic repairs before taking Nelson's body, which had been preserved in brandy, back to England for a state funeral, burial at St Paul's Cathedral and immortality in the pantheon of great Britons.

The ship underwent more extensive repair work at her Chatham Dock birthplace before being recommissioned in March 1808. On returning to the fleet she had a busy final four years of active service that included supporting

Sweden in its conflict with the Russian Empire, evacuating British troops from Spain and transporting reinforcements to Lisbon for Wellington's army in the Peninsular War.

In 1812 – nearly five decades after her launch – she was finally taken out of the front line and placed in reserve but, unlike most ships past their sell-by date, it was far from the end of her story.

Following refurbishment *Victory* took on the role of flagship for the Portsmouth Port Admiral and then, in 1831, came within a hair's breadth of being sent to the breaker's yard. Listed for disposal in that year she was granted an eleventh hour reprieve when her captain at Trafalgar, and now First Sea Lord, Thomas Hardy refused to sign the disposal order at his wife's insistence that the ship be spared.

Having been saved she served as a naval telegraphy training base from the late 1880s but was nearly sunk, in the most ironic fashion imaginable, by the Royal Navy. While being taken under tow to the scrapyard in 1903, ironclad turret ship HMS *Neptune* broke free and slammed straight into *Victory*, holing her below the waterline, and only emergency repairs prevented her sinking.

Had it not been for the launch of the 'Save Victory' campaign in 1921 it's likely she'd have deteriorated into such a state that disposal would be the only option. Fortunately for posterity, the campaign, bankrolled by shipping magnate James Caird, succeeded. In 1922 she was moved to a dry dock in Portsmouth, being in such poor condition she was no longer able to remain safely afloat.

Restoration started during the 1920s and returning her to her former glory for the nation has been an ongoing task ever since.

The Second World War showed just what *Victory* meant as a British status and propaganda symbol as the Luftwaffe claimed to have destroyed her during a bombing raid on Portsmouth Dockyard. While a bomb did land nearby, it only caused superficial damage and Germany – just like the Spanish and French – failed to get the better of *Victory*.

Today, *Victory* is one of the UK's top visitor attractions and welcomes on board some 350,000 people every year. She's the oldest commissioned ship in the world and, as part of the current Royal Navy fleet, maintenance costs are paid for out of the defence budget.

A living symbol of Pax Britannica, a monument to what Nelson and his navy achieved and a link to a moment that defined British, European and world history – *Victory* is arguably the greatest Royal Navy ship of them all.

A view of *Victory*'s wardroom.

The place on *Victory* where Nelson died.

A modern day view of *Victory*'s stern.

Victory at her berth in Portsmouth Dockyard, 2015.

CHAPTER TWO

HMS *Warrior*

HMS *Warrior*
9,284 tons
Built by Thames Ironworks & Shipbuilding Company, Blackwall, London
Served 1861–1883

Iron, steam and coal forged the Industrial Revolution that made Britain the world's greatest power – and all three were brought together in a ship that, when it took to the water in 1860, was larger, heavier and more powerful than any other vessel.

HMS *Warrior* was the first ever iron-hulled, armour-plated, steam-powered warship and, like HMS *Dreadnought* nearly half a century later, she changed the course of naval warfare.

Having routed the French fleet at Trafalgar, the Royal Navy was now indisputably master of the seas and the British government's trump card when it came to solving international disputes, protecting British interests or claiming new territory for the Crown.

While technology advanced at rapid pace throughout the nineteenth century – faster than at any other time in history – the ships that made up the Royal Navy fleet by the middle of the century were essentially just larger versions of the vessels that had fought at Trafalgar; bigger, but similar in appearance and design to HMS *Victory* and her sisters.

Steam power had been introduced into the British fleet as early as 1821, and it was generally agreed that steam rather than wind propulsion was the future, but with the early steam engines lacking the power of later models and being prone to mechanical defaults and breakdowns, ships were still equipped with masts and sails, running off a combination of steam and sail. Iron had also started to be used in ships, but their hulls were made of a mixture of wood and iron rather than exclusively from metal.

The Industrial Revolution led to the first industrial conflict – the Crimean War (1853–56), where mass-produced munitions and more modern military tactics were used both on land and at sea. Fleets of warships steamed in and out of the Baltic and Black seas to pound enemy positions and shipping, and, in one action, the Russians annihilated an Ottoman fleet made up entirely of wooden vessels. Yet, in a similar attack on three French vessels that had iron armour, the ships received more than fifty direct hits but only suffered superficial damage. The difference iron made convinced the Allied powers that armoured ships were undoubtedly the way forward.

They might have briefly been brothers in arms during the Crimean War, but Britain and France's relationship was still highly uneasy and war between them always a possibility, especially as both countries were now claiming large overseas empires in the same parts of the world. Keen to make up for British superiority in warship numbers, the French put their faith in technology by launching an armoured ship – *La Gloire* in 1859. With French foundries incapable of producing enough iron, *La Gloire* was designed as a wooden ship clad in iron 12 cm thick.

Up until now British naval planners saw their numerical superiority as the mainstay of the fleet's strength and were even reluctant to pursue new technologies due to the cost implications. Sir Baldwin Wake-Walker, Surveyor of the Navy, wrote in 1858:

> Although I have frequently stated that it is not in the interest of Great Britain – possessing as she does so large a navy – to adopt any important change in the construction of ships which might have the effect of rendering necessary the introduction of a new class of very costly vessels, until such a course is forced on us by the adoption by foreign powers of formidable ships of a novel character requiring similar ships to cope with them.

However, the launching of *La Gloire* was undoubtedly the adoption by a foreign power of a formidable ship and a direct challenge to British supremacy at sea.

Queen Victoria's Britain was never a country to back down from a challenge and the Admiralty set about responding with a new class of two ships that would render *La Gloire* immediately obsolete. With Britain's industrial capacity far outstripping that of France, producing enough iron for two ships wasn't a problem and the navy's Chief Constructor, Isaac Watts, got to work planning them.

Warrior, and her sister, HMS *Black Prince*, were designed from the outset as fully iron-hulled. Watts recognised that the wooden ships had reached their technological limit and that the size of vessels Britain now wanted could only be built out of iron. While the material they were made from changed, the shape and layout of the hull was the same as that used in wooden designs.

Another challenge *Warrior* posed for the engineers and shipbuilders was ensuring that the steam engine that would drive her could produce enough power to move such a large, heavy ship – far heavier than any wooden vessels steam engines had been previously been fitted in. Yet, typically for the era, it was a challenge they relished and rose to. Victoria's reign was a time of constant engineering and industrial progress in which Britain had boundless energy, limitless ambition and reached for the stars. *Warrior* was undoubtedly a shooting star – equipped with twenty-six 68-pounder guns, ten 110-pounders and four 40-pounders. The 110-pounder could fire shells 3,500 yards at a velocity of 340 metres a second and could be used against other ships or for shore bombardment.

On 11 May 1859, the Thames Ironworks Company won the contract to build the first ship of the class and its keel was laid on 25 May. Robert Napier & Sons at Govan on the Clyde was awarded the deal to build the second of the class which was laid down in October 1859.

Thames Ironworks had a long tradition of building iron ships, dating back to the opening of the yard in 1838. As well as constructing a variety of passenger and cargo vessels the company had previous warship experience, building a dispatch vessel for the Prussian navy and a frigate for the Royal Navy.

Although Thames Ironworks is now long gone, having closed in 1912, a football team formed by the works is still very much alive and kicking in the form of West Ham United. West Ham's club crest of two crossed hammers represent riveting hammers used in shipbuilding and the club's nickname 'The Hammers' refers to these tools that were used in the construction of *Warrior* and the many other ships built by Thames Ironworks.

More than 2,000 shipyard workers toiled day and night building *Warrior*, but, even with so many hands at work, the complexities of building an untried and untested ship led to inevitable delays and cost over-runs of the type all too familiar to those of us closely following the current construction progress of the new Queen Elizabeth Class aircraft carriers. In the end she cost the Admiralty £377,292 compared with her contract price of £264,664 – nearly double the price of a wooden steamship.

Warrior was launched on Saturday 29 December 1860 during a prolonged spell of freezing weather – the most severe to hit Britain for more than fifty years. The shipyard was covered in snow and it was so cold that the grease coating the slipway to aid her entry into the water had frozen solid. Despite the cold and biting wind a big crowd had turned out to witness the momentous occasion.

According to the *Liverpool Daily Post*:

A large concourse of persons assembled on Saturday to see the vessel launched from the Thames Ironworks, Blackwall, and at half-past two o'clock the hydraulic rams were set to work, but from the frozen state of the ways the *Warrior* was slow to move; she was ten minutes going the first two-feet six-inches. The steamers, which were attached to her for the purpose of assisting, then began to tell, and she slid easily and beautifully into the water, amidst loud cheering and was easily brought up by the hawsers made fast at her bows.

The Illustrated News also gave a detailed account of the launch, with the article conveying the sense of occasion and importance for a proud maritime nation.

> This formidable iron-cased frigate, the largest man-of-war ever built, and more than 1,500 tons larger than the largest vessel in the world after the *Great Eastern*, was safely launched into the river on Saturday. Although the day was one of those spitefully cold ones in which the frost often indulges when a coming south wind has given it notice to quit, the interest of the proceedings, both national and mechanical, attracted a very large crowd of visitors of high and low degree to Bow-creek. Most of the tall chimneys of the neighbourhood had been let out for the day, and were crowded by enthusiastic amateurs in shipbuilding, who could be seen in the aerial distance blowing their fingers and waiting patiently for the interesting moment when the signal gun was to announce the starting of the marine colossus. She looked splendid, agreeably disappointing those who had come down expecting to see a huge shapeless 'steam-ram,' bristling with iron plates, and who, to their surprise, found a noble ship on the finest lines, and exhibiting the most elegant proportions, looking, as the nautical men said, not half her tonnage, so symmetrical was she in form and outline. As the *Warrior* was an administrative child of the late government, it was to be expected that some of its members would be present to witness the debut of their off-spring. Accordingly Sir John Pakington was there, and General Peel, Colonel Knox, and some half dozen of their friends. The lords of the Admiralty present were Sir Richard Dundas and Captain Frederick.

The scale of the construction feat undertaken was so large that chairman of the Thames Ironworks & Shipbuilding Company, Peter Rolt, admitted having misgivings about the whole venture, saying, 'I often wondered how I mustered sufficient courage to undertake its construction.'

Launching the hull, however, is only the first part of building a ship – it's the fitting-out process that sees the installation of equipment and brings the vessel to life. The morning after her launch *Warrior* was moved to Victoria Docks for fitting-out.

In the age of iron and steam, fitting-out was much longer and more complicated than it had been with wooden ships. The steam engine, which had been made by the London-based engineering company John Penn & Sons, was partly assembled by the end of January and raising steam by March. A large part of the fitting-out process involved fixing the armour plates to the side of the ship, which proved a laborious and difficult task as each plate had to be bent to fit the curve of the hull before being grooved so it slotted tightly with the adjoining plates. More than 200 plates, weighing nearly 1,000 tons, were fitted in total.

Once *Warrior* was completed, the cost and time overruns were quickly forgotten as the Royal Navy now had the greatest ship in the world at its disposal and had seen off the French yet again without having to fire a single shot. *Warrior* showed that whatever France could design, develop and build, the British could always go one better due to the country's larger economy and far greater industrial output.

The reaction across the Channel to *Warrior's* arrival was predictably one of alarm, with one French naval officer commenting that, 'Should this ship meet our fleet, it will be as a black snake among rabbits!'

The Portsmouth-born great Victorian novelist Charles Dickens, as ever a wonder with words, described *Warrior* as 'a black vicious ugly customer as ever I saw, whale-like in size, and with as terrible a row of incisor teeth as ever closed on a French frigate'.

While the country and wider world were in awe of this new ship, for many of the crew serving in her it was just another ship. Jacky Fisher, who would later become First Sea Lord and play a key role in the development of HMS *Dreadnought* and lead the Royal Navy during the First World War, was a gunnery lieutenant on *Warrior* from 1863 to 1864. He later wrote that, at the time, those serving on the ship weren't aware of just how revolutionary she was, saying: 'It certainly was not appreciated that this, our first armour-clad ship of war, would cause a fundamental change in what had been in vogue for something like a thousand years.'

Commissioned into the Royal Navy's Channel Squadron on 1 August 1861 under the command of Captain the Honourable Arthur Auckland Cochrane, *Warrior* moved to her home port of Portsmouth the following month and began a series of trials to put the new ship and crew through their paces.

Operating under steam power alone *Warrior* could manage a top speed of just over 14 knots, 13 knots while using just sail power and more than 17 knots when operating with a combination of steam and sail – comfortably quicker than the 13 knots *La Gloire* could manage.

Life on board Royal Navy ships had improved markedly since the days of Nelson, but it was still, for the ratings if not the officers, what we'd regard today as a pretty grim and challenging existence.

Admiral George Ballard vividly described conditions on the battleship HMS *Resistance* in 1877, two years after he'd joined the navy. It was a description those serving on *Warrior*, or any other vessel at that time, would have recognised.

The bare bleakness of the mess deck had as little suggestion of physical ease as a prison cell. It was damp and chilly in a cold climate and damp and hot in the tropics. It was swept by searching draughts if the ports were open and nearly pitch dark if they were closed. Moreover the seaman was always in a crowd by day or night. His work and his leisure his eating, drinking, washing and sleeping were all in crowded surroundings.

Warrior's crew of around 700 comprised fifty officers, 100 petty officers and nearly 600 ratings and Royal Marines. The ratings were divided into thirty-four messes, each with around eighteen men squashed into the space between two guns. It was in this confined space that they'd eat, drink, spend time off watch and sling their hammocks when it was time to sleep.

While the ratings had to make do with basic mess decks, the surroundings for officers were far more salubrious. The captain's cabin was richly decorated and finely furnished, while the other officers all had their own individual cabins that they could transform into a home-from-home with their personal possessions.

Throughout 1862 *Warrior* welcomed visitors from the royal family, diplomatic service and the Admiralty, all of whom wanted to get up close and personal with the revolutionary new ship. Her maiden trip overseas saw her cruise to Lisbon for warmer water training and to promote relations between Britain and Portugal.

On returning to the UK she joined the rest of the Channel Squadron for cruises to Cobh in Ireland, now known as Cork, Milford Haven, Lisbon (again) and Gibraltar with her first foray into the Mediterranean.

In March 1863 *Warrior* was given the task of escorting the royal yacht bringing Princess Alexandra of Denmark to England for her marriage to the Prince of Wales, and she then toured the UK for twelve weeks, welcoming members of the public on board at various ports in the type of PR and goodwill exercise the Royal Navy is now renowned for. Throughout those three months an estimated 300,000 people visited the pride of their fleet to tour the ship and meet the crew.

In November 1864 she was decommissioned for a lengthy refit at Portsmouth Dockyard, during which she was refurbished and the 110-pounder guns, which had proved ineffectual and prone to faults, were removed and replaced with muzzle-loading guns. Watertight doors were also fitted and accesses cut through the iron bulkhead.

Recommissioned in 1867, *Warrior* attended a fleet review in honour of the Sultan of Turkey and the Khedive of Egypt and then rejoined the Channel Squadron under the command of Captain Henry Boys. Towards the end of the year she took on royal protection duties with a deployment to Osborne Bay to guard Queen Victoria while she stayed at Osborne House, her home on the Isle of Wight. With the Fenian Rising underway in Ireland, intelligence reports identified a possible threat to the monarch from Irish nationalists. As a way of showing her thanks for this protection the Queen made an informal visit to the ship.

Warrior's regal duties continued in 1868 when she escorted the royal yacht, HMY *Victoria and Albert II,* to Dublin for an official visit by the Prince of Wales before cruising to Scotland. While steaming off the Scottish coast she collided with the frigate HMS *Royal Oak*, losing her figurehead and jib boom. The figurehead – a carved wooden decoration attached to the ship's prow – was claimed as a prize by midshipmen on *Royal Oak*'s quarterdeck and Boys was court-martialled, though he was acquitted of blame for the collision.

Having ceased building wooden-hulled warships when *Warrior* and *Black Prince* entered service, the navy was now in desperate need of suitable docks for its new generation of ironclad vessels. To provide the West Indies station with adequate facilities, the Admiralty commissioned a floating dry dock that would be towed to Bermuda. *Northumberland*, *Agincourt* and the wooden paddle frigate *Terrible* towed the completed dock to the Portugese island of Madeira, where *Warrior* and her sister *Black Prince* took over the towing duties. The 2,700-mile voyage took more than three weeks as the ships' speed was greatly reduced by the load they were pulling, but they got the dock into place successfully and it remained in service off Bermuda until 1906.

After a refit to clean her hull and replace the figurehead lost in the collision with HMS *Royal Oak*, *Warrior* again became a member of the Channel Squadron. In March 1870, while returning from manoeuvres with the Mediterranean Fleet, she was present when HMS *Captain* capsized in bad weather off Cape Finisterre on the west coast of Spain with the loss of nearly 500 men. HMS *Captain* had been the navy's first attempt at building a turret ship where the vessel's guns are mounted in revolving turrets instead of a broadside arrangement. While *Captain* had been a catastrophic failure, the Admiralty learned much from her sinking and put these lessons into the building of HMS *Monarch* – the world's first successful turret ship. Turret ships had massive advantages over those with broadside guns as the guns could be pointed in any direction and had a more rapid rate of fire. *Warrior* had only been in service ten years but – such was the pace of change in the mid-Victorian era navy – she was already becoming obsolete.

Once a further extensive refit had been carried out between 1872 and 1875 *Warrior* became a reserve vessel and served as a guardship at Portland and then on the Firth of Clyde, stationed at Greenock. In March 1883 two of her masts were found to be rotten and, with no replacements available, she was decommissioned.

After decommissioning she was transferred back to her homeport of Portsmouth, where she was left in 'Rotten Row' as a floating hulk, although still officially classed as an armoured cruiser. Her masts and guns were stripped out when she was used as a depot ship for two years. In the early twentieth century she was renamed *Vernon III*, to free up the *Warrior* name for another new ship, and took on the role of providing steam and electricity to other hulks that formed part of the navy's torpedo training school.

Now of no further use to the Senior Service, she was put up for sale in 1924 but, with no buyers coming forward, she remained at Portsmouth for another five years before being towed to Pembroke Dock in south-west Wales, where she was used as a ship keeper's home and floating oil jetty. In half a century as an oil jetty, more than 5,000 ships refuelled alongside her and her hull, still strong and in good condition more than 100 years after she'd been launched, was a testament to the craftsmanship and quality of her build.

As early as the late 1960s there was talk of restoring *Warrior* to preserve part of Britain's illustrious naval heritage and act as an education resource for children. However at this stage she was still in use as an oil jetty – a role she continued in until the oil depot closed in 1978.

In the early 1960s Sir John Smith, then MP for the Cities of London and Westminster, had formed the Manifold Trust to restore threatened items of national heritage and when *Warrior* was no longer needed in 1978 the Trust agreed to underwrite the cost of restoration, which was estimated at between £4 million and £8 million. The ship was handed over to the Maritime Trust in 1979 and towed 800 miles around the coast to Hartlepool in the North East.

The Maritime Trust decided to take *Warrior* back to her 1861 condition and the restoration project was one of the biggest shipping restorations ever undertaken with 150 people working on it. Removing material added after her first commission, including a poop deck and concrete decking, alone took two years. To make the restoration as close to the original as possible, records, drawings and plans were scoured, including diaries and papers of those who'd served on the ship in 1861. An exact replica of the figurehead was even carved to replace the one lost in the 1960s, which itself had been a replacement following *Warrior*'s collision with the *Royal Oak*.

The only part of the restoration that did involve cheating slightly was that, with the engines and boilers considered too expensive to rebuild, replicas were rebuilt from sheet steel, although the replica engines, powered by electricity rather than steam, can rotate slowly to show visitors how they would've looked when in operation.

Warrior was in Hartlepool for more than seven years and so attached had the workforce and local people become to her that tears were shed in June 1987 when she was taken under tow to her new old home at Portsmouth, where a berth had been dredged and a jetty constructed for her arrival.

While in the English Channel, heading to Portsmouth, *Warrior* had a memorable encounter with the Royal Navy's newest ship – the Type 22 frigate HMS *London*. *London* signalled the message, 'The Navy's newest ironclad is in company with the oldest … I hope we look as good as you at your age.'

Her homecoming was of Prodigal Son proportions – thousands of people lined the dockyard and an armada of boats took to the water as she entered Portsmouth and moved towards her berth.

The ship is now the responsibility of the Warrior Preservation Trust – an independent charity that relies entirely on income raised from visitors, corporate hospitality, on-board weddings and donations to generate funds for *Warrior*'s upkeep and maintenance.

Visitors to the ship can see and try out practically everything from sitting at a mess table to taking a glance at the officers' wardroom and laying in a hammock – and the visitors flock in their hundreds of thousands every year.

Unlike the Americans, us Brits have never been very good at retaining old ships for heritage purposes, so, like HMS *Victory*, *Warrior* is an important link between the modern day and our illustrious naval past – a past that forged the world we now live in.

Warrior didn't fire a shot in anger and was only in active frontline service for a little over a decade. Yet her greatness is undisputed. After *Warrior* the Royal Navy, and the navies of all the other great powers, never built another wooden-hulled major warship. *Warrior* broke the mould and showed that ironcladding and steam power were the future. She maintained Royal Navy supremacy and is still as much of an attraction to the British public and conjures up the same feelings of pride in them today as she did when she came down the slipway at Thames Ironworks more than a century and a half ago.

Warrior at her berth in Portsmouth.

Above: Warrior's gun deck.

Opposite: Modern day *Warrior.*

CHAPTER THREE

HMS *Dreadnought*

HMS *Dreadnought*
18,400 tons
Built by HM Dockyard, Portsmouth
Served 1906–1919

Every so often new inventions or technologies come along and change the world. From the printing press to the internet and the motorcar to mobile phones, humankind's progress is made up of revolutionary developments that shape the course of history.

On a chilly, overcast day in February 1906, a steel hull slipped into the water at Portsmouth Dockyard. The ripples it created are still felt to this day.

HMS *Dreadnought* changed naval warfare so drastically that, with this one launch, every other single warship afloat was immediately out-dated.

Her introduction into service sparked a naval arms race with Germany that contributed to the world going to war in August 1914 and forced other navies to go back to the drawing-board to design and build fast, heavily armed and strongly armoured ships of their own.

With the exception of the first nuclear warhead, no one single military development has had such a colossal impact on the future of warfare and the course of world events as *Dreadnought*.

At the turn of the twentieth century, thanks to her vast navy, empire and unrivalled industrial output, Britain was still the pre-eminent world power, but a crisis of confidence was setting in.

Queen Victoria had been the very embodiment of British imperial rule. During her sixty-four-year reign, Britain's bounds were literally set wider still and wider as the country's riches and influence continued to grow. Not since the Romans had one empire held such sway and power over its territories and the wider world. For many, Queen Victoria was Britannia in the flesh so when she died in January 1901 at the age of eighty-one it was as though part of what Britain had been went with her.

While the aged queen lay on her deathbed, British forces in South Africa were suffering a series of setbacks and defeats at the hands of Dutch settlers in the Boer War. War had broken out in October 1899 over a disagreement between the Boers and British about voting rights in the Boer-controlled Transvaal. When the fighting started few expected anything other than a rapid victory for Britain's professional army – after all, they were only confronting poorly equipped Dutch farmers.

Unfortunately for Britain, the Dutch farmers proved to be a highly capable fighting force willing to mount ferocious resistance to defend their lands. Guerrilla-type raids by the Boers inflicted extensive casualties on a British Army performing well below par.

Eventually the British did prevail and the Boers surrendered in May 1902, but it was a victory that had taken so long and come at such cost that it left everyone from the people on the street to those in Whitehall's corridors of power questioning whether Britain was now in decline and losing her dominance.

The world's other great powers had been watching Britain's South African suffering with pleased interest – nowhere more so than in Germany. For decades the automatic assumption, both at home and abroad, was that, whether on land or sea, the British Empire was invincible. The Dutch farmers proved otherwise and it hadn't gone unnoticed.

Since Otto von Bismarck unified the German states into one nation in 1871 following Prussia's defeat of France in the Franco-Prussian War, Germany had enjoyed a meteoric industrial and economic rise and, along with the United States, was becoming a serious rival to Britain's world number one crown.

In 1871, British coal dominated world markets with production of 112 million tons a year. Germany, the world's second-largest producer of coal, mined 34 million tons. By 1890, German coal production was half of Britain's; by 1913 it was equal.

The figures for steel production also highlighted how Germany's industrial output was rapidly catching up with, then overtaking, Britain's. In 1890, Britain produced 3.6 million tons of steel a year, Germany around two-thirds that amount. By 1914, though, Germany (14 million tons) was producing twice as much as Britain.

The feeling in Germany was that this massive growth in the Reich's industrial and economic power and rapidly increasing population meant Germany needed an overseas empire. After all, Britain and France controlled vast swathes of territory across the globe and if the Germans were going to have equal, or even superior, status they needed the colonies to prove their standing. Securing colonial possessions and protecting the seas around them required substantial naval forces, so the Germans set about expanding their fleet. A large German navy would be a challenge to British supremacy both in European waters and in the sea-lanes around the new German Empire being carved out in central and southern Africa.

The German states, particularly Prussia, and then the unified Germany post-1871, had traditionally relied on a large army rather than sea power for security but, in Kaiser William II, Germany now had a leader with an enthusiasm for naval expansion. Given whom his maternal grandmother was, his desire to build a large fleet and forge a colonial empire was hardly surprising.

'I had a peculiar passion for the Navy,' William II wrote in his memoirs. 'It sprang to no small extent from my English blood …'

Frequent childhood trips to his grandmother Queen Victoria's seaside home, Osborne House, on the Isle of Wight fuelled his love for ships and the sea.

Just 5 miles across the Solent from the Isle of Wight lay the largest collection of warships in the world. The Royal Navy's base at Portsmouth captivated the imagination of the young German prince who was the son of Queen Victoria's eldest daughter, Vicky, and Prince Frederick William of Prussia.

> I often crossed over … to Portsmouth and saw all the classes of ships … and all the docks and shipyards … I climbed over the ship-of-the-line *Victory* … On the three-decker *St. Vincent* … gunnery practice was just taking place as I boarded her. I was permitted to take part … and told off as gunner No. 1 to serve a gun … I was not a little proud to have contributed my share to the deafening thunder of the broadside.

As the Jesuit boast goes, 'Give me the child for his first seven years, and I'll give you the man.' In William II's case these formative years of frequent exposure to the ships, officers and guns of the Royal Navy undoubtedly shaped his adult character and vision for Germany as a maritime power.

At a dinner with his uncle, King Edward VII, in June 1904 in Kiel, where every major ship of the German navy was anchored, William regaled the British monarch with a tale of how seeing the Royal Navy up close had sparked a desire to create Germany's own mighty fleet.

> When, as a little boy, I was allowed to visit Portsmouth and Plymouth hand in hand with kind aunts and friendly admirals, I admired the proud English ships in those two superb harbours, … Then there awoke in me the wish to build ships of my own like these someday, and when I was grown up to possess as fine a navy as the English.

For centuries Britain's prime enemy, and the country she was most prepared for war against, was France. From Agincourt to Trafalgar the story of the relationship between the two countries was one of hostility, suspicion and regular conflict. However, in 1904 the former adversaries suddenly put their differences aside and became close allies with the signing of the *entente cordiale*. The agreement settled many long-standing bones of contention. France recognised British control over Egypt, while Britain agreed to Morocco falling under the French sphere of influence. Binding themselves together in an alliance enabled both parties to focus on the new German threat while coming to each other's aid should the need arise.

With France now onside, the man Britain entrusted with maintaining Royal Navy supremacy over the growing fleet being assembled by Kaiser William II was First Sea Lord John 'Jacky' Fisher.

Born in humble surroundings in Ceylon (now Sri Lanka) to Sophie Fisher and army officer turned coffee planter Captain William Fisher in 1841, Fisher had joined the navy at the age of thirteen in 1854. Going on board Nelson's flagship, HMS *Victory*, to be assessed as to his suitability for service, the test consisted of writing out the Lord's Prayer and jumping naked over a chair in front of a doctor. Certified 'free from defect of speech, vision, rupture, or any other physical disability', he was immediately accepted.

Displaying enormous energy, talent and promise, Fisher enjoyed rapid promotion, seeing service in the navy's various stations around the world as well as regular postings ashore.

In 1881 he was appointed captain of HMS *Inflexible* – an ironclad ship that boasted heavier armour and bigger guns than any other vessel. This posting was considered the prize appointment in the navy and marked Fisher out as a man destined to reach the very top of his profession. The impact of a revolutionary battleship whose capabilities outstripped all competition wasn't lost on *Inflexible*'s skipper and would influence his thinking nearly a quarter of a century later in the creation of HMS *Dreadnought*.

After *Inflexible* had taken part in the 1882 Anglo-Egyptian war, bombarding the port of Alexandria, Fisher was struck down with a near fatal bout of dysentery that left his skin with a faintly yellowish complexion and added to the wholly baseless rumours that he was of Asian extraction and the son of a Malayan princess. In the less enlightened times that he lived, it led to the Admiralty's anti-Fisher faction dubbing him behind his back the 'Yellow Peril', 'The Malay' and 'that hobgoblin whose name is Fisher'.

Appointed a Rear-Admiral in 1890, then Vice-Admiral in 1896, at the turn of the twentieth century he was Commander of the Mediterranean Fleet – the most prestigious and well equipped of all the Royal Navy's fleets, responsible for an area from Gibraltar to the Suez Canal.

He then became Second Sea Lord, in charge of all personnel, in 1902 before finally reaching the highest office in the Royal Navy in 1904 with promotion to First Sea Lord.

He might have been sixty-three on taking up the post but Fisher possessed the energy, drive and enthusiasm of a man twenty years younger, regularly working sixteen-hour days including Sundays. On the Sabbath he'd head from the Admiralty to services at St Paul's Cathedral or Westminster Abbey, sometimes listening to up to four sermons through the course of the day, returning to his office with renewed zest and zeal.

As he never tired of telling anybody who'd listen, 'I entered the Navy penniless, friendless and forlorn. I have had to fight like hell and fighting like hell has made me what I am.'

In implementing his drastic reforms to overhaul and modernise the fleet he fought like hell from the moment he became First Sea Lord and was unbending in his determination to get what he wanted.

This is 'the house that Jack built,' he declared, 'so we must have no tinkering! No pandering to sentiment! No pity for anyone! We must be ruthless, relentless and remorseless! And we must therefore have The Scheme! The whole Scheme! And nothing but the Scheme!'

Older ships were paid off or put into reserve, the different stations around the world, including the West Indies and China stations, reorganised and cut to make them more modern and cost efficient. To counteract the growing German threat, battleships were withdrawn from the Mediterranean Fleet and reassigned to UK waters as part of the Home Fleet, which was renamed the Channel Fleet. The navy's gunnery skills were also scrutinised like never before, with Fisher insistent that gunnery had to be improved, while captains and senior officers whose ships and flotillas weren't up to scratch found their services dispensed with.

Right from the outset of his tenure as First Sea Lord, Fisher was certain Germany now proved the gravest threat to British security and all his reforms had this in mind. He said,

> The German Empire is the one power in political organization and in fighting strength and in fighting efficiency where one man (the kaiser) can press the button and be confident of hurling the whole force of the empire instantly, irresistibly and without warning on its enemy.

While reorganising his ships and removing both vessels and men who weren't up to the job formed part of his plans for modernising the navy, Fisher's flagship policy, as it were, was the creation of a mighty new battleship that was faster, stronger and more powerful than anything else afloat.

He pushed a decision through the Board of Admiralty that this new ship would be armed with ten 12-inch guns, carry 5,000 tons of armour and have a top speed of no less than 21 knots.

To make his dream a reality he convened a 'Committee on Designs' in January 1905 to produce detailed drawings and assist in the complex design process. Fisher appointed himself president of the committee and, even by the standards of his remarkable work ethic, pushed himself like never before to complete the task at hand in as short a timeframe as possible.

Dreadnought would be propelled through the water faster than any other warship thanks to the use of steam turbines rather than triple-expansion steam engines. She was the first battleship to be equipped with steam turbines that could push her to speeds steam engines would simply not be capable of reaching.

After much deliberation and tinkering with the designs, it was decided to mount her ten 12-inch guns in five twin turrets. The positioning of the guns meant she could deliver an eight-gun broadside that would rain down 6,800 pounds of ordnance on an enemy. Until *Dreadnought*, standard battleship armament in all major navies had consisted of four 12-inch guns, supported by a collection of smaller calibre weapons. Her ten 12-inchers made *Dreadnought* more than twice as powerful as any other warship.

Fisher outlined the thinking that had gone into *Dreadnought*'s design in a memorandum to the Cabinet:

> In designing this ship, the most powerfully arranged armament has been made the first consideration … Absolutely nothing has been allowed to stand in the way of the most nearly perfect power and scope of the guns… Being a battleship, she will have to fight other battleships. Having speed, she can choose the range at which she will fight.

In addition to the main armament of 12-inch guns, *Dreadnought* was also equipped with twenty-seven 12-pounder guns and five torpedo tubes.

As well as making the world marvel at her capabilities, Fisher wanted *Dreadnought* built in record time, so much of the steel that would make up the ship was ordered in standard sizes and piled up at Portsmouth's Royal Naval Dockyard, rather than using the traditional, slower, method of cutting each steel plate individually.

The build contract was awarded to Portsmouth as it had a reputation for building ships faster than any other British yard. Building in a naval dockyard at the home of the Royal Navy instead of a private yard also meant Fisher could keep a closer eye on construction and will the shipbuilders on to work as quickly as possible.

The keel was laid down on 2 October 1905 and, in a remarkable feat of unprecedented assembly speed, the hull was almost complete by New Year's Eve. As reward for such rapid construction Thomas Mitchell, the chief constructor of Portsmouth Dockyard, received a knighthood on 1 January 1906.

By February 1906 the hull was ready for launch and King Edward VII, a personal friend of Fisher, had been invited to do the honours.

Crowds of people, many arriving by specially laid on trains from London, swarmed to the dockyard to witness history in the making.

The Bishop of Winchester began the service to mark the launch with the 107th Psalm: 'They that go down to the sea in ships, that do business in great waters; these see the works of the Lord, and his wonders in the deep,' before blessing the hull and all who'd sail in her.

When the blocks keeping *Dreadnought* in place had been knocked away, the king swung a bottle of Australian wine at the bow. The bottle bounced back instead of smashing, so His Majesty swung again, shattering the glass at the second time of asking. He then cut the rope holding her on the slipway and she slid towards the water; slowly at first, then gathering speed.

The launch created headlines around the world and made journalists ponder what the introduction of such a ship might mean for international relations and the future of naval warfare.

The Times' Washington correspondent wrote:

> The building and launching of the *Dreadnought* have aroused interest here which is both professional and political. Nobody is more interested than the president [Theodore Roosevelt] to whom all things naval are of deep concern. That such a ship could have been built so quickly and so secretly astonishes naval experts.

Following the hull launch *Dreadnought* was moved to a fitting-out basin and brought to life with the installation of her engines, the fitting of turrets and guns and the addition of armour plating. Just as they had in constructing the hull, shipyard workers toiled around the clock to get the ship ready for sea trials as rapidly as was humanly possible.

On 1 October 1906 she slipped away from her moorings and headed out to sea for the first time for contractor's trials in the English Channel. Put through her paces at a variety of speeds and tested with all manner of manoeuvres, she performed just as well as the Admiralty, and Fisher in particular, had hoped and prayed she would. With gunnery trials proving equally successful *Dreadnought* was accepted into the Royal Navy on 11 December 1906. Such was her impact that she gave her name to all fast, heavily armoured battleships. From now on ships of this type would be referred to generically as dreadnoughts.

Her maiden deployment in early 1907 took her to Gibraltar and then into the Mediterranean before heading across the Atlantic to the West Indies, where further tests were carried out on her guns and machinery.

On her return to UK waters *Dreadnought* became the flagship of the Commander-in-Chief Home Fleet and embarked on something of a national tour, calling in at various British ports to give people across the country a glimpse of their mighty new ship. She then took pride of place at the annual naval review off Spithead and showed off her gunnery prowess to an audience including the king, queen and Prince of Wales.

For successfully introducing the fastest, most powerful battleship in the world into the British fleet and giving the pretenders desperate to knock Britain off her perch at the top of the global power tree a smack in the eye, Fisher could have been forgiven for thinking he'd now be feted as a national hero, a twentieth-century Nelson.

Yet the reaction from certain corners was strangely muted and in some cases downright hostile. The charge levelled at Fisher was that while Britain had enjoyed overwhelming dominance over other nations for more than a century in terms of the number of warships at the Royal Navy's disposal, he'd suddenly created a level playing field by rendering every other ship in the British Fleet, as well as all foreign opposition, obsolete.

A former First Sea Lord, Admiral of the Fleet Sir Frederick Richards, fired a broadside of his own at *Dreadnought*, complaining that, 'The whole British Fleet was … morally scrapped and labelled obsolete at the moment when it was at the zenith of its efficiency and equal not to two, but practically all the other navies of the world combined.'

Captain Reginald Bacon, naval assistant to the First Sea Lord and *Dreadnought*'s first skipper, vociferously defended the decision to press ahead with the revolutionary ship:

> Knowing as we did that the *Dreadnought* was the best type to build, should we knowingly have built the second-best type ship? … What would have been the verdict of the country if Germany had … built a *Dreadnought* while we were building Lord Nelsons, and then had forced a war on us and beaten our fleet? What would have been the verdict of the country if a subsequent inquiry had elicited the fact that those responsible at the Admiralty for the safety of the nation had deliberately recommended the building of second-class ships?

In building *Dreadnought*, Fisher and the Admiralty had unleashed a genie that could not be forced back into the bottle, so all the hollering and whinging in the world was futile.

Now Britain had created this new machine of war, it was up to Britain to produce more of them to maintain her numerical naval dominance. Even though Germany and the United States were rapidly catching up with Britain's industrial output, Britain was still far and away the world's biggest shipbuilder and had a much greater capacity for churning out dreadnoughts than any other rival.

In November 1905 the Unionist coalition government – made up of Conservatives and Liberal Unionists who'd split from the Liberal Party over the party's support of Irish home rule – pledged to lay down four dreadnoughts each year. Just as *Dreadnought*'s hull took to the water for the first time in February 1906, work started on three more dreadnought battleships – *Invincible*, *Inflexible* and *Indomitable* – at shipyards in Newcastle and on the Clyde. For all those complaining about Fisher creating a level playing field again, the British government did at least appear willing to build Britain back into the dominant position.

Yet the dreadnought building pledge made by the Unionist government proved meaningless as they were kicked out of office at the 1906 General Election and replaced in power by Henry Campbell-Bannerman's Liberal Party.

In their election campaign the Liberals had promised defence cuts and although they initially endorsed the construction of four new dreadnoughts this decision was quickly overturned. By May 1906 the new Chancellor of the Exchequer Herbert Asquith, who'd later become prime minister, was demanding the planned number of hulls be scaled back. From the four new ships planned each year one vessel was dropped for 1906, one for 1907 and two for 1908 – all at a time when Germany was dredging the Kiel Canal to make way for its own dreadnoughts and setting about an expansive building plan. Germany had laid down her first dreadnought in 1906, a further three in 1907 and four in 1909.

While the British government might be playing hardball with the number of ships it was prepared to build, the Admiralty told new First Lord of the Admiralty Reginald McKenna in December 1908 that it wanted six new dreadnoughts rather than four and that another six would be needed in 1910, as well as six more in 1911. Only by building in these numbers, the Admiralty argued, could Britain stay well ahead of Germany in the naval race that had now been sparked.

This position put the First Lord of the Admiralty at loggerheads with his fellow Cabinet members. Chancellor David Lloyd George, who'd moved into Number 11 Downing Street when Asquith replaced the ailing Campbell-Bannerman in Number 10, and President of the Board of Trade Winston Churchill were adamant that no more than four dreadnoughts would be approved.

Yet instead of yielding to the government's stance and agreeing to four, not six dreadnoughts, the Admiralty upped the stakes by increasing its request to eight vessels.

Deciding whether or not to give in to the Admiralty's demands split the Cabinet, with Churchill and Lloyd George sticking to their guns on four ships, the Foreign Secretary Sir Edward Grey wanting six and McKenna supporting the Admiralty in its calls for eight.

Eventually an agreement of sorts was reached in Cabinet whereby the government would ask for four dreadnoughts in 1909 and then seek authority to build a further four if Germany's rate of construction proved them necessary. While this four ships now, and maybe another four later, settled the matter around the Cabinet table it didn't in Parliament or the country at large.

On the Unionist side of the House, the Admiralty found willing cheerleaders in support of its dreadnought demands and the political battle lines were drawn over the battleships with the Unionist opposition, aided by the incensed editorials of the Unionist-supporting press, accusing the Liberal government of gravely weakening the Royal Navy and gambling with the British Empire's security by cutting the number of dreadnoughts.

'We want eight and we won't wait,' was the battle cry coined by the Unionist MP George Wyndham. It was a slogan that caught on and led to a public clamour that Asquith could not simply ignore.

The *Daily Telegraph* was scathing of the government's stance, stating that 'Since Nero fiddled there has never been a spectacle more strange, more lamentable, than the imperilling of the whole priceless heritage of centuries to balance a party budget.'

The Unionist leader Arthur Balfour articulated his side's position thus: 'In the opinion of this House, the declared policy of His Majesty's Government respecting the immediate provision of battleships of the newest type does not sufficiently secure the safety of the Empire.'

In the end, 'four ships now, four ships later' was voted through Parliament on party lines and Balfour's censure motion against it defeated by 353 votes to 135. They might have been beaten in the House but the Unionists did get their eight, and didn't have long to wait as it turned out.

In July 1909 it was revealed Austria and Italy were planning to build four dreadnoughts each, so the government gave the go-ahead to construct the contingency batch, plus another two funded by Australia and New Zealand. The two dominions both relied on the Royal Navy for protection so were keen to bolster the mother country's fleet.

Whereas before Britain had set having more ships than the next two biggest navies combined as its minimum standard, she now concentrated her efforts on her nearest rival rather than the two closest potential adversaries.

In March 1912 the new First Lord of the Admiralty Winston Churchill, now much keener to spend money on warships than he had been while President of the Board of Trade, stated that British policy was to maintain a 60 per cent superiority in dreadnoughts over Germany to show the kaiser that no matter how many new battleships he built, Britain would always build more.

In a letter to Fisher, Churchill said: 'Nothing in my opinion, would more surely dishearten Germany than the certain proof that as the result of all her present and prospective efforts, she will only be more hopelessly behind in 1920.'

With new sisters being built at a rapid rate around her, *Dreadnought* served as flagship of the Home Fleet from 1907–1911 and in 1910 was the victim of an elaborate and well-publicised hoax. Notorious hoaxer Horace de Vere Cole duped the Royal Navy into arranging for a party of Abyssinian royals to be given a tour of the ship. In reality the 'Abyssinians' were Cole and five friends decked out in African costumes with their faces blacked up. The hoax attracted much mirth and press coverage, much to the Senior Service's embarrassment.

In December 1912 she became flagship of the 4th Battle Squadron and was still in this role, based at Scapa Flow, at the outbreak of the First World War in August 1914. Ironically, while her construction had speeded up the journey to war and changed naval warfare forever, *Dreadnought* had a relatively quiet and uneventful time during the conflict. Her only significant action was the ramming and sinking of a German U-boat in the Pentland Firth near Scapa Flow in 1915. Having surfaced immediately in front of *Dreadnought* after firing a torpedo at HMS *Neptune*, the U-boat was chased down and sliced in two by *Dreadnought*, killing everyone on board. This gave her the distinction of being the only battleship to ever sink a submarine.

Undergoing a refit between 18 April and 22 June 1916, she missed the Battle of Jutland and when back in service was posted as flagship of the 3rd Battle Squadron based at Sheerness in Kent. With England's east coast being raided and bombarded by German ships on numerous occasions during the war, the Admiralty moved the 3rd Battle Squadron south from Rosyth on the Firth of Forth to Sheerness to reduce the distance they'd need to travel to intercept German raiding parties. Although built primarily for attacking rival battleships at sea, the only action *Dreadnought* saw at Sheerness was in an anti-aircraft role – firing at German aircraft flying over Kent to attack London.

In March 1918 she rejoined the Grand Fleet at Scapa Flow where once again she became, albeit briefly, flagship of the 4th Battle Squadron before another refit.

With Germany's surrender in November 1918 the need to maintain a mighty flotilla of battleships became far less pressing so *Dreadnought* fell victim to the peace dividend, being moved to Rosyth and paid off into reserve in February 1919. At the end of March 1920 she was stuck off the naval list and sold for scrap a year later.

Her service life might have been relatively short – just twelve years in total – and her war unexceptional, but no single ship before or since has ever left such a lasting legacy in its wake.

Dreadnought in harbour.

Dreadnought steaming in home waters.

Dreadnought's 12-inch guns just after firing. (Michael Pocock, MaritimeQuest.com)

Dreadnought is launched at Portsmouth Dockyard in February 1906. (Michael Pocock, MaritimeQuest.com)

CHAPTER FOUR

HMS *Warspite*

HMS *Warspite*
33,400 tons
Built by Devonport Dockyard
Served 1915–1945

HMS *Warspite*'s active service covered such an expanse of time that she fought at the Battle of Jutland and took part in the D-Day landings, nearly three decades later.

As a veteran of both world wars she saw action across the globe and won more battle honours than any other ship in Royal Navy history.

While *Warspite* served so illustriously for so long, she very nearly didn't last a year – suffering severe damage in the British fleet's mauling at Jutland in 1916.

Yet, like the young man who championed and gave the political sign-off to *Warspite* and her Queen Elizabeth Class sisters, she came back from adversity in the First World War to secure a place in the pantheon of immortality in the Second.

Winston Churchill had been appointed First Lord of the Admiralty in 1911 at the age of just thirty-seven. The post – as the politician in charge of the Royal Navy – placed him at the head of the most powerful organisation in the world and, never lacking in self-confidence and with a fondness for the limelight, it was an appointment he cherished, relished and threw himself into with customary energy.

By 1911 the naval arms race between Britain and Germany was reaching its peak as the rival sides pushed technology, and their finances, to the limit. The British still led the race with a larger number of battleships but while ten years earlier the Royal Navy had been lapping their Teutonic opponents, the Kriegsmarine was now closing the gap on the track heading for war.

Churchill's priority on taking up residence at Admiralty House was to widen the gap again and re-establish his fleet's overwhelming dominance over Germany and any other potential rival who might threaten British mastery of the seas.

To do so he knew a major statement – of both his and Britain's intentions – was needed. A statement that would show the kaiser and the rest of the world that challenging the Royal Navy was foolish and futile.

Following the successful introduction of 13.5-inch guns aboard the King George V, Orion and Iron Duke Class super-dreadnoughts, Churchill decided that his statement would be a new class of battleships equipped with 15-inch guns. Unlike their predecessors, these battleships would be driven by oil instead of coal and would be the most powerful vessels ever to put to sea.

While coal had served the Royal Navy well for decades, coaling a ship was time consuming and labour intensive. The saving in weight from burning oil rather than coal would mean heavier guns could be carried and greater propulsion power generated. It also removed the need for regular replenishment stops at coaling stations.

As Churchill later recalled in *The World Crisis*, his account of the First World War and its build up, the decision to press ahead with the Queen Elizabeth Class caused him much apprehension:

> No such thing as a modern 15-inch gun existed … None had ever been made. The advance to the 13.5-inch had in itself been a great stride. Its power was greater; its accuracy was greater; its life was much longer. Could the British designers repeat this triumph on a still larger scale and in a still more intense form? I hardly remember ever to have had more anxiety about any administrative decision than this.

As well as the uncertainty about whether the Queen Elizabeths were pushing the boundaries of naval architecture too far, Churchill also agonised over the expense of the project and questioned himself over whether the ships really were needed, noting in *The World Crisis*: 'Was it wanted? Was it the right thing to make? Was its tactical value sufficient to justify the increase in costs and all the changes in design?'

By nailing his colours to the Queen Elizabeth Class mast, Churchill was taking a leap of faith into untried and untested waters – and putting his political reputation on the line. For nobody was even certain 15-inch guns – the largest calibre ever proposed for a battleship – could be successfully mounted on and fired from a floating vessel. A full load of 900 15-inch shells to feed the guns would alone weigh 1,000 tons and the turrets 550 tons each.

Working to Churchill's brief, the Admiralty's ship designers and the Elswick Ordnance Company were tested like never before but they delivered for their political master by designing the Queen Elizabeths – which were initially intended to be a class of six super-dreadnoughts.

Work began on HMS *Queen Elizabeth* and HMS *Warspite* in the autumn of 1912 and *Valiant*, *Barham* and *Malaya* were all laid down in 1913. *Agincourt*, the sixth ship in the class, was cancelled in August 1914 as it was argued she wouldn't be ready until after the war was over and would therefore serve little purpose.

Given the political capital the First Lord of the Admiralty had invested in the Queen Elizabeth project, it was with much relief that, on 26 November 1913, he watched *Warspite* slide down the Devonport Dockyard slipway and into the waters of the River Tamar, cheered on her way by a 30,000-strong crowd who spontaneously broke into a rousing rendition of 'Rule Britannia'.

In addition to the eight revolutionary 15-inch guns mounted in four twin turrets, *Warspite* was armed with fourteen 6-inch guns, two anti-aircraft weapons, four saluting guns and four submerged torpedo tubes.

Between her launch and commissioning in March 1915, the conflict many had predicted for years, between Europe's great powers, had broken out in August 1914, so *Warspite* was on a war footing from the moment she joined the fleet.

For her first crew, serving in such a state-of-the-art ship was a novel and exciting experience. One of her 15-inch gunners recalled:

> I immediately felt I would be at home and happy. This happened to be so throughout the time I served in her. She was commanded by Captain Phillpotts and we were in good hands being commanded by such a gallant gentleman. The ship's 15-inch guns were a much bigger armament than any battleship had carried before. It was something new and at the same time something strange. The ship herself soon proved everything that was expected of her, having been built in Plymouth and starting from Plymouth with a West Country captain.

During acceptance trials in the Irish Sea that August, on her way to join Britain's Grand Fleet at its wartime base at Scapa Flow, *Warspite*'s pace and successful firing of the 15-inch guns fully vindicated Churchill's decision to press ahead with the Queen Elizabeth Class. In a ninety-minute dash she maintained a speed of 24.5 knots and the guns turned out to be everything the Admiralty hoped they would – as well as packing more of a punch than smaller guns, the 15-inchers proved more accurate over longer ranges. Churchill noted that the first time he witnessed the firing of a 15-inch gun he felt 'delivered from a great peril'.

While the introduction of *Warspite* and the other Queen Elizabeth battleships added unprecedented power to the British fleet, her active career got off to a less than glorious start. Visiting Rosyth in mid-September, she ran aground off Dunbar having been led down the wrong channel by escorting destroyers and then, three months later during a Grand Fleet exercise, she collided with *Barham*, causing such extensive damage to her sister ship's bows that it was initially feared *Barham* would sink.

After repair work at Plymouth following the collision, *Warspite* returned to Scapa Flow on Christmas Eve before being sent to a new temporary home at Rosyth. From there, the Admiralty felt she was in a more suitable location to intercept German attacks on Britain's east coast.

In the early months of the war the kaiser's fleet had bombarded a series of towns – Hartlepool, Whitby, Scarborough and Great Yarmouth – injuring nearly 600 people and killing more than 130 including women and children. For a population brought up to believe in the supremacy of the Royal Navy, which they'd been led to believe would protect them and their Empire against all foes anywhere in the world, the British people were starting to feel let down. If His Majesty's ships couldn't stop the Germans shelling their homes then perhaps the Senior Service wasn't the mighty, invincible force they thought it was.

For the navy, this loss of public confidence was a sobering experience and one the Admiralty was keen to rectify. What it needed was a set-piece, full on confrontation at sea that would rekindle the spirit of Nelson and Trafalgar. That chance came in May 1916 when the German High Seas Fleet lured the Grand Fleet into battle in a bid to break the naval blockade the British held the Germans under, with Britain controlling the English Channel and access to the North Sea between the United Kingdom and Norway.

The German plan was to use Vice Admiral Franz Hipper's fast scouting group of five battlecruisers to draw Admiral Sir David Beatty's battlecruiser squadrons into the path of the main German fleet and the submarines that lay in wait over their anticipated path.

Unbeknown to Germany's naval command, the British had picked up from signal intercepts that a major operation was likely, so the Grand Fleet steamed east in large numbers, ready for a titanic clash of steel and the mightiest ever battle at sea.

As part of Beatty's battlecruiser force, *Warspite* was despatched alongside her sisters *Barham*, *Valiant* and *Malaya* for what would be the ultimate test for the 15-inch guns Churchill had championed.

At 3.58 p.m. on 31 May 1916, the five German battlecruisers – *Von Der Tann*, *Moltke*, *Seydlitz*, *Derfflinger* and *Lützow* – fired the opening salvos of the Battle of Jutland off the north-west coast of Denmark.

HMS *Indefatigable* was hit and destroyed by the first barrage before *Warspite* and *Malaya* aimed their guns and blasted back at *Von Der Tann*. The first shots fell short but hit home when corrected, with *Warspite* landing a strike on *Von Der Tann*. Then all hell broke loose as *Warspite* turned and the rival fleets closed in and ferociously pounded each other. Commander Humphrey Walwyn, who was on board *Warspite*, recalled:

Very soon after the turn I saw on the starboard quarter the whole of the High Sea Fleet: at least I saw masts, funnels and an endless ripple of orange flashes all down the line, how many I didn't try and count as we were getting well strafed at this time, but I remember counting up to eight … The noise of their shells over and short was deafening … Felt one or two very heavy shakes but didn't think very much of it at the time and it never occurred to me that we were being hit.

While German shells were raining down on her, *Warspite* gave as good as she got and hit back.

Commander Walwyn recalled seeing the devastating impact of her 15-inch guns striking their target: 'I distinctly saw two of our salvos hit the leading German battleship,' he said. 'Sheets of yellow flame went right over her mastheads and she looked red fore and aft like a burning haystack; I know we hit her hard.'

The strikes on *Warspite* started to claim casualties, with many suffering burns from the now blazing upper deck and superstructure. At this stage it looked as though *Warspite*'s battle to stay afloat was going to be in vain – and it probably would have been were it not for the heroic efforts of her crew in fighting fires and plugging holes in the ship's hull with everything from hammocks to wooden furniture. Walwyn recalled:

> Men everywhere were simply splendid, and all so cheery and although I confess it was mighty unpleasant and unnerving I myself had plenty to do, but for those who merely had to wait it must have been a thousand times worse … The noise was so perfectly appalling, and you couldn't hear at all between decks and the worst of it was knowing nothing.

The damage from a shell striking her port-wing engine room jammed *Warspite*'s steering as she performed a hard turn to avoid colliding with *Valiant* and *Malaya*. The jamming left her unable to manoeuvre in a straight line. All she could do was turn in circles but Captain Phillpotts ordered her to continue turning, figuring it was better to keep moving than become a stationary target for the High Seas Fleet.

While *Warspite* now appeared something of a lame duck, her damaged rudder provided salvation for the critically damaged armoured cruiser *Warrior*. Like a big cat switching its attention to juicier prey, the kaiser's ships that had been blasting away at *Warrior* suddenly turned their attentions to the Queen Elizabeth Class battleship, thereby diverting fire away from *Warrior*.

Miraculously, despite numerous further hits, *Warspite* remained afloat and engineers even managed to rectify the steering fault, allowing her to escape.

Sub Lieutenant Vaux could not believe the ship didn't sink as he saw:

> … about four or five German dreadnoughts firing at us for about twenty minutes and hitting us about once a minute. Why we weren't sunk is a perfect mystery, personally, after seeing the *Defence* [an armoured cruiser] go. I thought we should go every minute. I was simply soaked with salt water from the splashes of their shells that fell short.

Fortunately for *Warspite*, the dreadnought HMS *Agincourt* arrived on the scene just in time. The broadside she fired dispersed the German ships and enabled *Warspite* to slip away.

Finally able to fully assess the damage to their ship, *Warspite*'s crew were shocked by what they saw and knew that, while the battle against the High Seas Fleet might now be over, the battle to stay afloat and get back to Rosyth had just begun. Commander Walwyn went on:

> When the ship got out of the range of enemy guns, the order was passed to us saying crews could stand easy but remain in the vicinity of the turrets … This meant you could come out and walk around the top of the turret. When I got out I was amazed. There were fires under the bridge where the lifebelts had caught fire, there were fires in the 6-inch battery where the cordite was alight. The funnels were holed – every boat in the ship had a hole in it. The ship looked really bad.

With various compartments flooded and her engine rooms in danger of being swamped with water, the battered dreadnought could only limp along at 6 knots.

For Surgeon Lieutenant Ellis, the damage to *Warspite* was a sickening blow:

> Only a few hours before she had been one of the cleanest and smartest looking ships in the fleet, her decks spotlessly white, and her light grey paint, freshly put on only recently, gleaming everywhere in the sunshine … Now her decks were filthy, littered with debris and in places torn up by shells, one of the quarterdeck ladders had been blown away, her funnels had ragged holes in them.

Nearing Rosyth, she had to run the gauntlet of German U-boats lurking off the coast. Operating at such reduced speed meant outrunning the German underwater menace wasn't an option. The wounded *Warspite* made easy pickings and, sure enough, a submarine detected her and went in for the kill. Somehow all three torpedoes missed.

If the crew were expecting a tumultusou reception for having survived the battle and successfully crawled home they were to be sorely disappointed.

News was just filtering back to the UK that the great Trafalgar-type victory the nation wanted and expected had been nothing of the sort. Instead, the supposedly invincible Royal Navy had lost more ships than the Germans and, even worse, its tactics and conduct in the battle was open to question.

The battlecruisers *Indefatigable*, *Queen Mary* and *Invincible* had all been lost and nearly 6,800 British sailors killed, compared to 3,000 German dead.

After the shelling of towns on the east coast, the lack of a decisive outcome at Jutland was too much for the British people to stomach and many were openly hostile and critical of the navy's performance.

Bill Fell, a midshipman on *Warspite*, noted the resentment of the locals in Rosyth:

I don't think we gave any thought to how we would be received at home … But as we passed up the Firth of Forth and under the bridge, all the railway people were lined along it. To our dismay they shouted 'Cowards! Cowards, you ran away' and chucked lumps of coal at us. We were received at Rosyth with very, very great disapproval by the local people. They were all in mourning black hats and black arm-bands. They all felt the Grand Fleet had suffered complete defeat and that some ships, like the *Warspite*, had run away.

Warspite lost fourteen men killed and eighteen wounded – a miraculously low number considering the damage she sustained. When transferred to a dry dock it was found she'd been holed around 150 times and the repair work required was so extensive it took two months to complete, even at the speeded up wartime work rate.

After completion of repairs and sea trials in the Firth of Forth, she returned to Scapa Flow at the end of July 1916 to rejoin the fifth battle squadron and had a largely uneventful time until the end of the war.

Ten days after the Armistice on 11 November 1918, *Warspite* and the rest of the Grand Fleet put to sea for a rendezvous with the surrendering German fleet and escorted them to internment at Scapa where, with no neutral country prepared to provide the Kriegsmarine with a harbour, the German commanders scuttled their ships seven months later.

In 1919 she joined the second battle squadron as part of the newly formed Atlantic Fleet that replaced the disbanded Grand Fleet and made regular cruises to the Mediterranean.

A major two-year refit that started in 1924 increased *Warspite*'s protective armour and saw the addition of more modern, powerful anti-aircraft guns. Once the refit was complete she was transferred to Malta, where she became flagship of the Commander-in-Chief Mediterranean Fleet before rejoining the Atlantic Fleet in 1930.

A second major refit between March 1934 and March 1937 effectively gave the Admiralty a new battleship, such was the extent of the work to her engines, internal machinery and weapon systems. Two cranes and an aircraft hangar with space for four planes were also fitted. After the extensive overhaul *Warspite* returned to the Mediterranean and was in Valletta harbour when Britain and France declared war on Hitler's Germany in September 1939 following the German invasion of Poland.

As it had at the outbreak of hostilities in 1914, the Royal Navy's Home Fleet took up wartime anchorage at Scapa Flow. Little more than a month into the war a German U-boat penetrated the British defences around Scapa and sank the battleship *Royal Oak* at her moorings – an action that claimed the lives of more than 800 men.

With the Home Fleet now in need of reinforcements to make up for the loss of the *Royal Oak*, *Warspite* was recalled from the Mediterranean and prepared for battle with Germany, twenty-three years after she'd so nearly gone down at Jutland.

In April 1940, German forces invaded Denmark and Norway and seized the Norwegian Arctic port of Narvik, from where Swedish iron ore was exported to aid the Nazi war effort. Such was the strategic value of Narvik that around ten German warships were based there so a flotilla of British destroyers was sent to confront them, sinking two of the Nazi vessels on 10 April.

The remaining eight German ships now effectively remained trapped at Narvik due to a lack of fuel and battle damage. Having so many vulnerable German vessels in one place at the same time proved too tempting an opportunity for the Royal Navy to miss, so *Warspite* and nine destroyers were despatched to finish them off.

Warspite's Fairey Swordfish floatplane led the way, attacking and sinking an enemy U-boat and then acting as a spotter to direct British fire on to the German flotilla. *Warspite*'s 15-inch guns blasted away at ships, shore batteries and clifftop gun emplacements.

Arthur White, a junior rating on HMS *Eskimo*, witnessed *Warspite*'s devastating barrage:

> There was a rushing noise like a bloomin' express train. It put the fear of Christ in us. We wondered what the hell it was. All of a sudden this destroyer and the cliff behind us blew up.

Having sunk all the German ships, *Warspite* and her escorting destroyers took part in further shore bombardments in Norwegian waters before steaming back to Scapa Flow.

From the cold waters of the Arctic, the warmer seas of the Mediterranean were *Warspite*'s next destination and she arrived in Alexandria – the Mediterranean Fleet's wartime home – just as Italy was about to enter the war on the Axis side.

In July 1940 the British and Italian navies clashed around 30 miles off the toe of Italy at the Battle of Calabria.

Warspite, leading the British battleships into action, opened fire at lead Italian ship *Giulio Cesare* and achieved one of the longest range gunnery hits in history by a moving ship on a moving target – hitting Italy's dreadnought at a distance of 24 kilometres.

Admiral Andrew Cunningham, the Royal Navy's Commander-in-Chief Mediterranean, witnessed the outstanding feat of gunnery from the 15-inch guns that had been championed all those years earlier by the man who was now Britain's new prime minister. He said:

> The *Warspite*'s shooting was consistently good … I had been watching the great splashes of our 15-inch salvoes straddling the target when at 4 p.m. I saw the great

orange-coloured flash of a heavy explosion at the base of the enemy flagship's funnels. It was followed by an upheaval of smoke, and I knew that she had been heavily hit at the prodigious range of 13 miles.

The long-range hit struck *Giulio Cesare*'s rear deck, killing sixty-six sailors, and shut down half of the ship's boilers, which led to her withdrawing from action. The battle ended indecisively with neither fleet managing to sink any rival ships, although both sides claimed to have had the upper hand.

With Calabria finishing in stalemate, a more conclusive encounter was required if the Royal Navy was to inflict serious damage on its Italian foes and remove them as a force from the Mediterranean. It was decided an air attack by torpedo bombers against the Italian fleet at anchor in its main port of Taranto would be the most suitable way to achieve this aim.

Warspite was tasked with providing cover to HMS *Illustrious*, the aircraft carrier launching the planes carrying out the successful torpedo strike that sank one Italian battleship and badly damaged two others.

Victory at Taranto meant the British now really did have the upper hand in the naval war in the Mediterranean and this dominance was extended even further at the Battle of Cape Matapan off the south-west coast of Greece in March 1941.

Following up air attacks launched by the carrier HMS *Formidable*, the three Queen Elizabeth Class sisters, *Warspite*, *Valiant* and *Barham*, pursued the Italians and sank two heavy cruisers and two destroyers at point-blank range. Lieutenant T. Campling, a gunnery officer on HMS *Formidable*, described the sinkings:

> Within seconds there was an ear splitting roar as the 15-inch guns of our battleships opened fire at almost point-blank range. The result was immediate and devastating. I vividly recall seeing a complete turret of the leading ship disappear over the side. Masses of flame soon enveloped all three ships.

From success at Matapan, *Warspite* suffered tragedy while operating in an anti-aircraft role during the German invasion of Crete in May 1941. A 500 lb Luftwaffe bomb slammed into her, killing thirty-eight men. Fortunately she could still move under her own steam and limped back to Alexandria, although the damage was so extensive she had to be transferred to the United States for repairs to be carried out as there weren't the facilities in Egypt to undertake such a major job.

Following the completion of repairs, *Warspite* was sent to the Indian Ocean as part of the Eastern Fleet at a particularly low point for British fortunes in the Far East. Singapore had recently fallen to the Japanese and 100,000 British and Empire troops taken prisoner – the largest capitulation in British military history. The battleships *Repulse* and the brand new *Prince of Wales* had also been lost to Japanese air attacks, so it seemed in early 1942 that the Land of the Rising Sun was on an unstoppable rise, while the sun was setting on British power in a region where it had exercised imperial control for more than a century.

In August 1942 *Warspite* participated on a simulated attack on the Andaman Islands to distract Japanese forces while the United States prepared to retake Guadalcanal, and she was involved in covering the landings during the Allied operation to oust the Vichy French from Madagascar.

Having returned to the UK in early 1943 she was then once again sent to her now highly familiar surroundings of the Mediterranean, where she bombarded Italian coastal batteries in support of the Allied invasion of Sicily before reprising the role she'd undertaken in 1919 when she escorted surrendered German ships to Scapa Flow. This time she led elements of the Italian fleet to internment in Malta to prevent them falling into German hands when Italy signed an armistice with the Allies.

During Allied operations in Italy *Warspite* was regularly called up to the gun line to bombard enemy positions, and it was while undertaking a bombardment assignment that she was again attacked and damaged by the Luftwaffe. Three Dornier bombers armed with primitive guided missiles went in for the kill. One of the missiles struck and gouged a twenty-foot hole in her hull, killing nine and wounding fourteen of those on board.

She was patched up at Rosyth, where her 6-inch guns were removed and concrete used to fill in the hole left by the German missile, just in time to take part in the D-Day landings.

With age creeping up on her and unable to undertake the longer voyages and battles that had once been her bread and butter, *Warspite* was now primarily used as a floating artillery platform.

Germany's Atlantic Wall defences were made up of an extensive system of coastal fortifications stretching from northern Norway to the south of France that were in place to repulse an Allied invasion, wherever in occupied Europe it might come.

Softening up these defences would be crucial to the success of the Allied landings and, in HMS *Warspite*, the Allies had a tool ideally suited to this type of operation.

The vast Allied armada that assembled in the English Channel in the early hours of 6 June 1944 crept quietly towards the Normandy coast, and the first sound to break the silence was the thunderous roar of the old ship's trusty 15-inchers.

Ordinary Seaman Ronald Martin was on board *Warspite* for the D-Day landings. He recalled:

As twilight approached there were 6,000 or 7,000 ships off the south of the Isle of Wight, which was a magnificent sight … The soldiers, poor devils, were so sea-sick, they were leaning over the sides of the landing-craft. Led by a fleet of minesweepers we crept very slowly towards Gold Beach at the eastern end of the beach-head. We arrived just before dawn was breaking and at 05.30, or just before, we sighted what we thought were torpedo E-boats coming out from Le Havre to attack us. I opened fire with the port armament. Almost at the same time the 15-inch guns opened up against our first shore battery. So the *Warspite* lays claim to firing the first shots of the invasion.

One unfortunate, or foolish, German Panzer commander made the mistake of lighting up while sitting on top of his tank, and the glow from his cigarette was spotted by reconnaissance aircraft and gave away the position of the armoured column hiding in a wood.

Ronald Martin picks up the story:

We pin-pointed the target and gave an order unique in naval gunnery history of '15-inch, fifty rounds, rapid fire. Commence.' That was almost impossible because of the time it took to load one-ton shells, follow their pointers and so on. However, there was a method in that madness because the first broadside fell slap in the middle of the wood and then as the succeeding ones started, because of the human error and the time lag, the shells were chasing the Germans as they started to run away. We got reports back from the aircraft that it was very successful. We could knock a chap off his bicycle at 20 miles.

We then ran out of 15-inch ammunition after nearly 400 rounds. So we went back to Portsmouth where we loaded ammunition all night. Halfway through the night we got a signal saying that the Americans had got themselves into all sorts of bother down on Omaha beach. Their bombarding wasn't going very well and they had specifically asked for *Warspite* to go down and sort them out. We went straight there to bomb the German troop emplacements with the aid of aircraft and spotters. They would communicate by radio to give the map coordinates. Then we went back to our original bombarding positions and carried on bombarding.

It was somehow appropriate that the guns Churchill had staked his early political career on played a role in the operation that liberated occupied Europe and paved the way for the ultimate victory Churchill had promised his people when it looked but a forlorn hope in the dark days of 1940.

After D-Day, and with the Allies firmly established ashore in western Europe, *Warspite* continued to bombard enemy positions during the Battle for Brest in September 1944 and then supported an Anglo-Canadian operation to clear German gun emplacements and strongholds around the Belgian port of Antwerp.

After Antwerp her guns fell terminally silent. In February 1945 the Admiralty confirmed *Warspite* would be paid off into reserve as she was too old and battle-worn to continue in service without a costly major rebuild that, with the war coming to an end, could not be justified.

In death though, just as in life, she refused to go without a fight. While being taken under tow to Scotland by two tugs for scrapping in April 1947 she broke free of her tow line in rough seas off Lands End and ran aground. She finally beached off St Michael's Mount, where she remained on view as a wreck until she was completely cut up and removed in the mid-1950s.

As one newspaper put it in her obituary when she ran aground: 'Her oldest loves, the wind and the sea, have helped the old lady of the fleet to cheat the executioner. She is ashore, apparently for good …'

It was a fitting end for the most battle-hardened ship to ever serve the Royal Navy.

Warspite entering Valetta harbour in Malta during the Spanish Civil War.

Warspite engaged in shore bombardment off the coast of Normandy.

Warspite runs aground off the Cornish coast on her way to the breaker's yard. (Michael Pocock, MaritimeQuest.com)

CHAPTER FIVE

HMS *Hood*

HMS *Hood*
47,000 tons
Served 1920–1941
Built by John Brown & Company, Clydebank

Hood was the largest, fastest ship afloat, pride of the Royal Navy and the very epitome of Britain's power at sea.

Yet within just 90 seconds she was gone, blown apart by the *Bismarck* and sent to the bottom of the icy Denmark Strait along with all but three of her 1,418 crew.

By 1941 the British were accustomed to the war bringing bad news. Dunkirk, Norway, Greece – all had ended in ignominious defeat and retreat in the face of the German juggernaut, blitzkrieging its way across much of Europe and North Africa.

Even Winston Churchill, replacing the hapless Neville Chamberlain at 10 Downing Street in May 1940, failed to bring about a reversal in Britain's fortunes. While his oratory and fighting spirit stirred hearts and stiffened resolve it had little impact where it really mattered. As the Labour front-bencher Aneurin Bevan wryly commented, 'The prime minister wins debate after debate and loses battle after battle.'

Yet even after nearly two years of losses, setbacks and failure, the sinking of HMS *Hood* was a devastating blow to British prestige and, perhaps, a sign that winning the war was within Germany's grasp - for *Hood* was far more than just a ship.

While Britain eventually prevailed in the First World War, it was in many ways a hollow triumph. A whole generation perished in the mud of Flanders, Ypres and the Somme and it had been a less than glorious episode for the world's most respected and feared fighting force. In 1914 the British people expected their navy to fight and win a magnificent victory.

However, Britain's Grand Fleet and the admirals and politicians commanding it were so keen to avoid defeat that they ended up too afraid to attempt delivering a knockout blow against their German foes. Instead of putting to sea early on to do battle, the bulk of the Royal Navy spent much of the war holed up in its anchorage at Scapa Flow, content to keep a wary eye on the Kreigsmarine from afar.

When the rival fleets finally did engage at the Battle of Jutland in 1916 it was no repeat of Trafalgar. The British lost more ships and limped back to Scapa to lick their wounds, but did at least inflict enough damage to keep the Germans in port for the rest of the war.

At the end of hostilities the Royal Navy was still by far the biggest in the world – and in theory victorious – but the failure to deliver the decisive blow harmed its standing. It was a standing that needed rebuilding, and with *Hood* the Admiralty thought it had just the solution.

Although ordered during the war, *Hood* wasn't launched until three months before the Armistice. Taking on board the lessons painfully learned at Jutland, her design was radically revised while under construction at John Brown & Company's Clydebank shipyard.

Originally intended as the first of four Admiral Class battlecruisers that were planned to rival the German Mackensen Class, none of her sisters – or Germany's Mackensens – ever saw the light of day, for the end of the war meant they were all dispensed with before sliding down a slipway.

Hood's build, though, was so far advanced as the conflict reached its close that finishing her made sense economically, as well as providing a welcome opportunity to create an unrivalled post-war status symbol.

Named in honour of the great Georgian admiral, and Nelson's contemporary, Viscount Samuel Hood, it was a title deliberately chosen to remind Britain, and the world, of the Royal Navy's most celebrated period and to restore a reputation tarnished by the First World War.

At Jutland, Britain's Grand Fleet had proved vulnerable with three battlecruisers blowing up under German shellfire due to inadequate armour, so providing *Hood* with greater protection became a priority.

At the cost of slightly decreasing her speed, an extra 5,000 tonnes of armour was added above and below the waterline, although failing to install stronger deck armour would have tragic consequences more than two decades later.

In the early twentieth century the River Clyde was the beating heart of the shipbuilding world and the industry established Glasgow as the British Empire's second city.

'Clydebuilt' became such a hallmark of quality that around a quarter of all ships constructed anywhere on earth during the late Victorian and early Edwardian eras were made on this one Scottish river.

So when the navy wanted the best they placed their orders on the Clyde – and no shipbuilder came with a finer reputation than John Brown & Company of Clydebank.

Employing 7,000 men in the shipyard – and a further 3,000 in its nearby engine works – by 1916 the company had already given birth to some of the greatest merchant and military vessels of the age, including the Cunard Line's flagships *Lusitania* and *Aquitania*.

Despite John Brown's wealth of experience in building trailblazing ships, *Hood* was of such a size and scale that it tested the skills of all involved.

Built to outrun and outgun any other warship afloat, her specifications were mindboggling. More than 260 metres long and over 30 metres wide, her turbines produced 144,000 shaft horsepower that could propel her through the water at 32 knots.

Armed with eight 15-inch and twelve 5.5-inch guns, torpedoes and a fearsome array of anti-aircraft weaponry, *Hood* packed the heaviest of heavyweight punches.

The 15-inch guns could blast a one-tonne shell – that shot out of the muzzle at more than 1,600 mph – 17 miles.

Unlike today, when ships are usually constructed in pre-fabricated blocks, *Hood* was fashioned from individual steel plates that were cut and bent to the required shape and size before being riveted together around her keel and frame.

Following her launch in August 1918 the fitting-out process began. This saw the installation of items including funnels, conning tower, bridge, armour belt and guns.

While she was put together and brought to life on the Clyde, thousands of men and women across the United Kingdom played a part in *Hood*'s construction, from smelters and forgers in Sheffield producing her steel to the workers at Vickers in Barrow-in-Furness creating the 15-inch guns and turrets.

Being built during wartime meant less attention and fanfare was given to the ship's progress and launch, partly to prevent the enemy finding out much about her, but also because with thousands still dying daily on the Western Front singing her praises would have felt inappropriate.

The relative lack of column inches, and therefore limited wider public awareness of this mighty new ship taking shape, made her thrust into the limelight after the end of the war all the more dramatic.

As the *Dumbarton Herald* noted in January 1919, 'Very few people have hitherto been aware of her existence, and yet the *Hood* is a far more wonderful vessel than the Hush! Hush! Ships which made such a sensation [*Glorious*, *Courageous* and *Furious*].'

Not only did people marvel at her size and firepower, but at her elegant appearance, for *Hood*'s lines and shape made her one of the finest looking warships ever built – a beauty that added to the ship's symbolism and aura.

With orders flowing in at John Brown & Company to replace merchant shipping lost during the war, *Hood* was moved from west to east Scotland in January 1920 to finish fitting-out at Rosyth.

Her sheer scale was highlighted by the fact that Rosyth had one of only three dry docks in Britain – the others being at Liverpool and Plymouth – large enough to accommodate her.

Once fitting-out was complete and sea trials passed, *Hood* became a fully commissioned Royal Navy ship in May 1920 and was based at Devonport. The final build cost came in at just over £6 million – more than £200 million in today's money.

Right from the start of her career *Hood* and her crew received special treatment that was the envy of much of the rest of the fleet. A posting in *Hood* was seen as a highly coveted, plum appointment that carried a special prestige. Even though she'd only been in service for six months, and didn't see action during the First World War, in November 1920 her marine detachment was selected to provide the guard of honour at the burial of the Unknown Warrior on Armistice Day in Westminster Abbey.

This role at the centre of national life continued for the next twenty years and the new world order forged out of the First World War's ashes played right into *Hood*'s hands.

The Washington Naval Treaty – signed by Britain, France, Japan, Italy and the United States in 1922 – put strict limitations on the building of battleships, battlecruisers and aircraft carriers in an attempt to prevent an arms race that could lead to another global conflict. So with the major powers not having any new capital vessels on the horizon, *Hood* remained unchallenged as the mightiest ship afloat. It was a status Britain and the Royal Navy revelled in.

With peace now restored to the world, it soon became clear that *Hood*'s role would be to act as an ambassador and 47,000-tonne diplomat, protecting Britain's interests and flying the flag for the British Empire, now bigger than ever with German overseas territories turned pink on the map following Germany's First World War defeat.

In the term made famous by US President Ronald Reagan some sixty years later, *Hood* provided peace through strength and for this to be effective she had to be frequently seen and shown off.

On commission she became flagship of the Battlecruiser Squadron of the Atlantic Fleet under the command of Rear Admiral Sir Roger Keyes, and wasted no time taking on her ambassadorial role.

Hood's first deployment was to the Baltic Sea. Joined by the battlecruiser *Tiger* and nine destroyers, she steamed north to show the fledgling Soviet Union what it would be up against if the new socialist state threatened the sovereignty of Britain's allies in the Baltic states.

After the Baltic she headed for a goodwill tour of Scandinavia which included a memorial service off the coast of Denmark for the dead of Jutland – the battle that had radically altered *Hood*'s design while on John Brown's slipway. Respects to their fallen comrades paid, the fun began and *Hood* received a rapturous welcome from her Scandinavian hosts.

Lieutenant Commander Douglas Fairbairn paints a vivid picture of the ship's reception while meandering through the Norwegian fjords:

At each little village we passed were crowds of cheering Norwegians, some of whom even swam out towards the ship. At every white flagstaff among the trees the Norwegian flag was flying, and dipped in salute to us as we went by: this continued for a whole hour. It was a wonderful welcome and to those ashore the mighty *Hood*, the largest warship in the world, winding her way through these landlocked waters must have been a magnificent sight.

With Scandinavia suitably impressed, *Hood* headed for home and then the warmer waters of the Mediterranean in 1921 to train with the Mediterranean Fleet and show off her power in this key strategic region for the British Empire.

Further warm water deployments followed with a trip to Rio de Janeiro to celebrate the centenary of Brazil's independence alongside the Brazilian, American and Japanese navies before visiting the West Indies.

These early diplomatic missions, however, were just a warm-up for the main event that was to follow in 1923/24.

The Cruise of the Special Service Squadron, or the Empire Cruise as it became known, was planned as an unabashed show of Royal Navy might to remind the dominions of their reliance on British sea power, and was the greatest circumnavigation of the globe undertaken by the Senior Service since Commodore George Anson's voyage around the world in the 1740s.

The squadron, made up of battlecruisers *Hood* and *Repulse* and the light cruisers *Delhi*, *Danae*, *Dragon*, *Dauntless* and *Adelaide*, left Devonport on 27 November 1923 and would visit numerous colonies and dependencies as well as the United States in a ten-month deployment.

Sierra Leone was the first port of call in the African leg of the journey, which also took in Cape Town, Durban, Zanzibar and Dar es Salaam. Each stop involved much pomp and pageantry ashore and gave Britain's colonial subjects the opportunity to take an up-close look at the mother country's most famous and prized naval asset.

After Africa, the Indian Ocean and Far East beckoned. The squadron called in at Ceylon, Malaysia and the recently expanded British naval base at Singapore that had become the Royal Navy's home in the Far East.

The biggest public interest in the cruise came in Australia and New Zealand, where hundreds of thousands turned out to see the Special Service Squadron. In Melbourne alone more than 200,000 people boarded *Hood* and just as many had to be turned away.

Lieutenant Geoffrey Wells recalled:

We remained at Melbourne for seven days and enormous crowds besieged us. Numbers unprecedented percolated into every place in the ship. Women fainted in the gangways which almost gave beneath the weight. To get ashore became a feat of no mean skill and elbow power. The *Repulse* also suffered. The little boys of the crowd fared best, they could squeeze where their sisters could not and could deal more rapidly with the ladders.

As well as showing off their ships, the British sailors attended numerous official functions and fundraisers and took on their Australian hosts at sport with a cricket match against Melbourne Cricket Club at the MCG.

On leaving Australasia, the squadron headed for the Pacific, then the eastern seaboards of Canada and the United States, before undertaking the most dramatic part of the trip – navigating the Panama Canal.

Hood's size meant she literally had just inches to spare on either side passing through the canal's locks and the spectacle of this mighty ship squeezing down the waterway connecting the Pacific and Atlantic oceans drew crowds all along the route.

Returning to Britain following further stops in Jamaica, Nova Scotia, Quebec and Newfoundland, the statistics associated with the Empire Cruise were staggering. The Special Service Squadron burned 110,000 tonnes of fuel oil – more than twice the weight of HMS *Hood* herself – and some 2 million visitors came on board the ships during the 38,000-mile voyage.

For a navy and country whose reputation had suffered badly in the First World War, the cruise was seen as a British imperial triumph that restored pride and honour and showed the world that Britannia still ruled the waves.

Throughout the rest of the 1920s *Hood* made regular tours around the British Isles and spent each spring visiting Spain and exercising in the Mediterranean, keeping a watchful eye on the region that was the strategic lifeline of the British Empire.

After nine years of service she underwent a major refit between May 1929 and March 1931, but world events would lead to a stormy reintroduction to the fleet that rocked the foundations of Britain's political and military establishment.

The 1929 Wall Street Crash sparked a Great Depression that saw currencies devalued, rampant high inflation and mass unemployment across the globe. In a bid to slow the UK's snowballing budget deficit, the government proposed drastic public sector cuts that included slashing the wages of military personnel.

It was announced that all men in the navy being paid on the higher 1919 scales would be placed on the lower 1925 rate – a move that would affect pretty much all chief and petty officers and three out of four other rates, while officers stood to lose less.

Hood had resumed service and was at anchor with the Atlantic Fleet at Invergordon in the Cromarty Firth, north of Inverness, preparing for manoeuvres when news came through of the wage reductions. In response the crews of *Hood*, *Nelson*, *Valiant* and *Rodney* refused to put to sea and so began the Invergordon Mutiny – one of the few military strikes in British history.

For two days crews on board *Hood* and the other three battleships declined to carry out all but essential duties, cheering crowds massed on the decks and officers who tried to issue orders were ignored and ridiculed.

Leading Seaman Sam Wheat was in no doubt who the strikers held responsible:

Austen Chamberlain [First Lord of the Admiralty] was the bloke – they'd have crucified him if they could've got him, believe you me. Because he thought that none of the navy had any intelligence at all. To take a shilling a day from each one was, to put it mildly, bloody silly. I mean, they thought that the chaps were unintelligent and this is really what caused it. I think they'd have got away with it if they'd said 'Well, we're going to reduce your pay by 5 per cent' or whatever it is … But to do that made them look like nincompoops as though they didn't know anything at all.

With the world watching, sailors threatening to damage machinery and panic on the London Stock Exchange leading to a run on the pound, the government and Admiralty rapidly drew up a solution. Cabinet agreed to the Admiralty's recommendation that ratings on the old rate of pay remain on that scale, while taking a 10 per cent cut in line with the rest of the service.

This settlement was agreed but a number of strike organisers were jailed and 200 sailors discharged from service for their part in the mutiny. It was perhaps ironic that the 'Mighty *Hood*', pride of the Royal Navy, had played a key role in such a shameful episode.

The mutiny behind her, *Hood* led the Battlecruiser Squadron on a cruise to the Caribbean in early 1932 before heading to her old winter stomping ground of the Mediterranean later in the year.

With the Nazis sweeping to power in Germany in 1933 the peace that had looked set to last in 1918 now appeared much less certain. But, for now, the tyranny that would be responsible for *Hood*'s destruction was appeased by Britain's political establishment and the great warship ordered to continue her diplomatic duties, rendered ignorant to the impending crisis by the heads in the sand approach of the politicians controlling her.

It wasn't just the Germans now presenting themselves as potential adversaries either. With Mussolini's Italy intent on carving out a new Roman empire in Abyssinia and Japanese expansionism threatening British interests in the Far East, Britain faced the prospect of fighting on multiple fronts. But for now, the peace lasted.

Embarrassingly for the Royal Navy, *Hood*'s most notable action of the mid-1930s was an all-British affair. While en route to Gibraltar in 1935 she was rammed by fellow battlecruiser *Renown* and had to limp to the Rock with a damaged hull and propeller for patch-up work before returning to Portsmouth for permanent repairs. Squadron commander Rear Admiral Sidney Bailey, *Hood*'s captain Thomas Tower and *Renown*'s skipper H. R. Sawbridge were all court-martialled over the incident. Although Tower was acquitted of any wrongdoing, the Admiralty partially pointed the finger of blame at Bailey, criticising him for ambiguous signals during the manoeuvre that led to the collision.

As flagship of the Royal Navy, *Hood* was at the centre of King George V's Silver Jubilee Fleet Review off Spithead and then during the Spanish Civil War in 1937 she escorted British merchant ships into Bilbao, despite Nationalist attempts to blockade the port.

With war looming ever closer, she was withdrawn from the Mediterranean and overhauled in Portsmouth between January and August 1939 before being transferred to the Home Fleet at Scapa Flow.

Now more than twenty years old, time was beginning to catch up with the mighty *Hood* but the planned major refit and modernisation, which would have included the addition of more adequate deck armour, didn't materialise as she was needed for wartime service.

Hood spent the first few months of war protecting convoys in the North Atlantic before being sent south to play a leading role in one of the most controversial episodes in Royal Navy history.

France had the second biggest navy in Europe so after the French surrender in June 1940 Britain faced the nightmare scenario of her defeated ally's large fleet falling into German hands and being used against the United Kingdom.

To avoid this disastrous outcome Churchill gave Admiral Francois Darlan, Commander-in-Chief of the French Navy, a three-optioned ultimatum: sail with the Royal Navy and continue the fight against Germany; sail to UK ports and hand over the ships to Britain for the duration of the war; or head to French ports in the West Indies to put them out of German reach. If all three choices were refused then the Royal Navy would have no option but to sink the French fleet at its Algerian anchorage in Mers-el-Kebir.

As the War Cabinet awaited Darlan's response, the British ships – led by *Hood* and the aircraft carrier *Ark Royal* – moved up to the gun line. The ultimatum having been refused, the Royal Navy opened fire with one of the most ferocious naval bombardments ever seen.

Ted Briggs was a Boy Signalman aboard *Hood*. He recalls:

Just as I turned round to watch, the guns of the *Resolution* and *Valiant* roared in murderous hair-trigger reaction. Then came the ting-ting of our firing bell. Seconds later my ears felt as though they had been sandwiched between two manhole covers. The concussion of the *Hood*'s eight 15-inch guns, screaming in horrendous harmony, shook the flag deck violently.

Stoker First Class Vernon Coles witnessed the horror unfold from the destroyer HMS *Faulknor*:

At 17.55 we opened fire. It was a sad irony. We were not attacking the Germans or Italians, but the Royal Navy's oldest enemy and our twentieth century ally. The whole fleet was going across and ranging. What a bombardment! I had never seen anything like it. One of our destroyers out on the starboard wing had got so close inshore that she was coming under the range of their 9-inch gun, so the *Hood* just trained her guns to fire at the hill, just below a big fort which was where the firing was coming from. The fort came tumbling down because the blast had undermined its foundations. The French battleship *Dunkerque* was right under a dockyard crane and the *Hood* had to destroy it before she could get at the *Dunkerque*. Her first broadside hit the crane, it was just like a matelot dropping. The second salvo hit the *Dunkerque*. We were firing from a distance of 7 or 8 miles, which for a 15-inch gun is point-blank range.

The barrage left 1,300 French sailors dead, including more than 1,000 in the battleship *Bretagne* that was blown apart by *Hood*, *Valiant* and *Resolution*. The Axis powers – and many in France – accused Britain of war crimes for this ruthless, but ultimately necessary, course of action. The resolve of Churchill and his war cabinet in taking this most difficult of decisions convinced the United States the British had the will to resist Nazism whatever the cost, as well as taking battleships that could have been used in an attempted German invasion of the United Kingdom out of Hitler's hands.

Returning to home waters, *Hood*, along with *Nelson* and *Rodney*, was dispatched south from Scapa to Rosyth on the Firth of Forth to be closer to the anticipated German invasion fleet that – thanks to the RAF's victory in the Battle of Britain – never came.

The threat of invasion diminished, *Hood* resumed her convoy escort duties before undergoing yet another brief and inadequate refit between January and March 1941.

On her return to service she now had two new fearsome enemies to contend with – the mighty Bismarck Class battleships *Bismarck* and *Tirpitz*, the largest and most heavily armed ships Germany had ever put to sea.

Far more modern than *Hood* and with superior armour protection – particularly against plunging fire – the new German sisters posed a huge menace to the Royal Navy and the merchant convoys Britain relied on for her very survival.

AB Len Williams, who'd served on *Hood* between 1936 and the spring of 1941, feared for his former ship:

As an ex-member of *Hood*'s crew I can recall numerous discussions we had in our mess about a possible meeting with either *Bismarck* or her sister *Tirpitz*. We were not at all happy about such a prospect. We knew our weakness and the risks of not having an armoured deck. We had the speed, yes, and we had the gun power; but we did not have our armour in the right place!

In the Denmark Strait, Williams' fears became a nightmarish reality.

When *Bismarck* set out on her maiden operational deployment *Hood* and the newly commissioned *Prince of Wales* led a British force to prevent her getting into the Atlantic shipping lanes and attacking convoys transporting invaluable supplies of food and war materials to Britain.

Having raised anchor in occupied Norway, *Bismarck* and the heavy cruiser *Prinz Eugen* headed north with the aim of making for the gap between the west coast of Iceland and eastern Greenland and breaking out into the North Atlantic before the British Home Fleet, steaming from Scapa Flow, could intercept them.

On the evening of 23 May the heavy cruisers HMS *Norfolk* and *Suffolk*, patrolling the Denmark Strait, spotted the two German ships and tracked them with radar throughout the night.

The next morning the British force of *Hood*, *Prince of Wales* and six destroyers was in place to attack as the Germans passed south-west of Iceland.

Hood opened fire at a range of about 15 miles and steamed towards the enemy at full speed. Vice Admiral Lancelot Holland, commanding the British force, figured that closing the distance better protected *Hood* from plunging fire due to the incoming shell trajectory being flatter the closer they got.

While *Prince of Wales* hit *Bismarck*, the shells didn't cause serious damage and she shot back at the British with devastating consequences.

The first impact hit *Hood*'s boat deck and started a fire, but it was the next salvo that wiped out the pride of the Royal Navy.

Struck somewhere near her main mast, the lack of deck armour meant the shell pierced *Hood*'s magazine, causing a catastrophic explosion that sent a gigantic fireball into the sky and tore the ship in two.

Three years to build, twenty-one years of service and she was gone in less than two minutes.

In *Lost Voices of the Royal Navy*, Leading Sick Birth Attendant Sam Wood recounts the clear view he had from another ship in the British force of the *Hood*'s destruction:

> The leading seaman with me said, 'Christ, look how close the firing is getting to the *Hood*'. As I looked out, suddenly the *Hood* exploded. She was just one pall of black smoke which blacked us out. Time seemed to stand still. I just watched in horror. The bows pointed out of this smoke, just the bows tilted up and then this whole apparition slid out of sight, all in slow motion, just slid slowly away. I couldn't believe it. The *Hood* had gone.

Ex-crew member Leonard Williams, who'd feared for *Hood*'s fate, pictured the death and destruction wrought on his former shipmates:

In a tremendous flash, a split second of searing time *Hood* was gone, rendering all our efforts null and void. After serving for four and a half years in the ship I knew every compartment, nut and bolt in her. I can almost picture the terrible scene between decks when that fatal shell struck. The gigantic sheets of golden cordite flame sweeping through the narrow corridors and passages, incinerating everything in its path. The terrific hot blast, the bursting open of the armoured hull under the colossal pressure; and, finally, the merciful avalanche of the cold sea, cleansing the charred and riven wreck, and bringing peace to those gallant souls I knew so well. On more than one occasion I have dreamed this scene and have returned to consciousness with the thought that 'There, but for the grace of God went I.'

The cataclysmic explosion and fire that ripped through the ship so instantaneously killed 99.8 per cent of those on board. Somehow three men – Ordinary Signalman Ted Briggs, Able Seaman Bob Tilburn and Midshipman William Dundas – survived.

Ted Briggs, the last of the survivors, who died at the age of eighty-five in 2008, felt himself being pulled under with the mighty ship but was miraculously propelled to the surface and spared:

When I came to the surface I was on her port side, something I have never been able to figure out! She was about 50 yards away, her bows vertical with the water and B turret just going under. I turned and swam as best I could in water 4 inches thick with oil and managed to get on one of the small rafts she carried, of which there were a large number floating around. When I turned again she had gone and there was a fire on the water where her bows had been. Over on the other side I saw Dundas and Tilburn on similar rafts. There was not another soul to be seen. We hand-paddled towards each other and held on to one another's rafts until our hands became too numb to do so.

Able Seaman Bob Tilburn recalled the sickening damage inflicted on his shipmates' bodies:

The next shell came aft and the ship shook like mad. I was next to the gun shield so I was protected from the blast, but one of my mates was killed and the other had his side cut open by a splinter. It opened him up like a butcher and all his innards were coming out. Bits of bodies were falling over the deck and one hit me on the legs. I thought, 'I'm going to be sick' so I got up and went to the ship's side to throw up. Then I looked up and saw the bows coming out of the water. The *Hood* was turning over. I jumped on to the forecastle which was nearly under water and started to strip off: tin hat, gas mask, duffel coat and all the rest. By then the water had reached me and I was swimming.

As if to rub salt into the already agonizing wound to Britain's pride and prestige, the loss of *Hood* – the symbol of British imperial power who'd travelled around the colonies in the 1924 cruise – came on Empire Day.

Reeling from her loss, the Royal Navy, furious with grief and desperate for revenge, chased *Bismarck* from the Denmark Strait and relentlessly pursued the German flagship as she ploughed through the North Atlantic towards safety in occupied France.

Swordfish planes flying from the aircraft carrier *Ark Royal* launched torpedoes that destroyed her rudder and steering gear, leaving her effectively a lame, albeit still highly dangerous, duck.

On the morning of 27 May the battleships *Rodney* and *King George V* and heavy cruisers *Norfolk* and *Dorsetshire* closed in for the kill, pounding *Bismarck* until she was left so badly damaged the crew had no option but to scuttle her to save the pride of the Kriegsmarine falling into British hands.

Hood was avenged, but nothing could ease the pain of her loss. For when *Hood* went down part of the Britain she stood for and represented went with her too.

Hood at Scapa Flow in 1940. (Michael Pocock, MaritimeQuest.com)

A view of *Hood*'s stern. Her seaplane is visible towards the front of the photo.

Hood seen at Station Pier, Melbourne, Australia, during the world cruise. (Michael Pocock, MaritimeQuest.com)

CHAPTER SIX

HMS *Rodney*

HMS *Rodney*
33,700 tons
Built by Cammell Laird, Birkenhead
Served 1927–1946

She didn't serve the dish cold, but in large explosions, burning flame and twisted steel. As *Rodney* closed in to near point-blank range on the floundering German behemoth she was hell-bent on revenge.

Revenge for what this 42,000-ton monster had done three days earlier, many hundreds of miles to the north, in the Denmark Strait.

When the *Bismarck* sank their venerated *Hood*, the Royal Navy pursued her seeking the type of vengeance usually reserved for the pages of the Old Testament.

The whole fury of the fleet was directed at relentlessly hunting, then destroying the Nazi battleship – and it was the 16-inch guns of HMS *Rodney* that delivered the death-blow.

In serving up the revenge the British people not only wanted, but needed, after the sickening loss of *Hood*, *Rodney* became a national treasure overnight and a Royal Navy great.

Like HMS *Hood*, *Rodney* was named in honour of a Georgian admiral. George Brydges Rodney (1718–92) was a man of contradictions. On the one hand considered by fellow officers and subordinates to be arrogant, vain, selfish and a compulsive gambler obsessed with money, he was also one of the most able naval commanders of his generation, popular and frequently kind-hearted.

Ironically, given the role *Rodney* played in avenging *Hood*'s destruction, George Brydges Rodney and Samuel Hood were on less than cordial terms for much of their careers. While serving as Rodney's second-in-command, Hood accused his superior of sacrificing the Senior Service's interest to his own profit and of lacking in energy in pursuit of the French fleet at the Battle of Saintes in the Caribbean in 1782.

Rodney's father, Henry Rodney, had served in the army during the War of the Spanish Succession and on leaving the forces lost all his money with an investment in the South Sea Company, so when George Brydges arrived in 1718 the family were impoverished. This experience of growing up in poverty thanks to a bad investment more than likely shaped Rodney's cavalier attitude towards money.

Despite their financial difficulties the Rodneys were well connected and these connections enabled George to attend Harrow School before joining the Royal Navy at the age of fourteen.

Promoted from midshipman to lieutenant in 1739, he served in various ships in the North Sea, English Channel and Mediterranean, becoming a post captain at the age of just twenty-five in 1743. In early 1746 Rodney was appointed master and commander of the 60-gun HMS *Eagle* and took the ship into action against the French at the second Battle of Cape Finisterre a year later. While Rodney was in command *Eagle* captured sixteen enemy ships – a feat that saw Rodney scoop £15,000 in prize money, equivalent to nearly £2million today.

On top of the financial remuneration for his success in *Eagle*, he was rewarded with promotion to governor and commander-in-chief of the British colony of Newfoundland with the rank of commodore. The governorship of Newfoundland at the time usually went to a naval officer due to the rich fish stocks in the colony's waters. Each summer a large British fishing fleet sailed for Newfoundland to catch cod and the Royal Navy's task was to protect the trawlers from potential attack by French warships sailing from France's nearby colony of Quebec.

Rodney's time in Newfoundland was largely uneventful and in 1752 he turned his attentions to politics, being elected MP for Saltash, a now abolished parliamentary constituency in Cornwall.

In the latter half of the eighteenth century the Royal Navy saw almost continuous active service and Rodney took part in a host of campaigns during the Seven Years' War and the American War of Independence, with his naval career culminating in the 1782 destruction of a French fleet off Dominica at the Battle of Saintes which established his status as a national naval hero.

Prior to the creation of the subject of this chapter there had been five other HMS *Rodney*s named in George Brydges' honour – two of them in his own lifetime and three in the nineteenth century.

The *Rodney* we're concerning ourselves with here was lucky to ever see the light of day due to an international agreement to limit warship construction in the wake of the First World War.

As we've already seen, the naval arms race between Britain and Germany had been a contributing factor to the outbreak of war in 1914; a war that up to that point had been the darkest, costliest man-made episode in human history. Given the colossal carnage, bloodshed and loss of a generation that it wrought, it's hardly surprising the powers that suffered so extensively were keen to build a new world where future conflicts of this scale would never be repeated. Not for nothing was the First World War dubbed 'the war to end all wars'.

When peace came on the eleventh hour of the eleventh day of the eleventh month in 1918, the victors – in truth a hollow title – very quickly got together around the negotiating table to thrash out agreements for what they hoped to build from the war's ashes.

The main outcome of 1919's Paris Peace Conference was the establishment of the League of Nations. The League of Nations, essentially an ultimately flawed forerunner to the United Nations, was founded in January 1920 as an international organisation with the aim of promoting and maintaining world peace through collective security, disarmament and the settling of international disputes through arbitration instead of on the battlefield.

While President Woodrow Wilson championed the creation of the League of Nations, the United States Senate voted against American involvement in the fledgling organisation over fears it would lead to loss of American sovereignty and drag US servicemen into wars that weren't of their making. Without the economic, diplomatic and military clout of the emerging new world power behind it, the League was struggling from the start.

America might have opted out of the League of Nations but it was at the forefront of a naval agreement signed by itself, Britain, France, Italy and Japan in 1922. The Washington Naval Treaty put limits on warship construction to avoid naval expansion programmes that might lead directly or indirectly to another world war.

California's Republican Senator Hiram Johnson exalted that: 'War may be banished from the earth more nearly by disarmament than by any other agency or in any other manner.'

Today the idea of a Republican Senator being in favour of disarmament sounds an alien concept, but for people of all political creeds in the early 1920s disarmament seemed like the only option to prevent the world falling into an abyss.

Although the Washington Treaty didn't disarm the navies of the respective powers, it did reign in the force they could unleash.

The terms of the agreement decreed that Britain and the United States were permitted capital ships with a total tonnage of 525,000 tons, Japan 315,000 tons and France and Italy 175,000 tons apiece. It also placed limits on aircraft carriers, implemented a ten-year pause or 'holiday' on the construction of new battleships and put in place a restriction that no single ship was to exceed 35,000 tons or mount guns larger than 16-inches.

The number of other categories of warships – cruisers, destroyers and submarines – was not limited although they were not permitted to weigh in at any more than 10,000 tons.

For the British it was an uneasy compromise born out of necessity. While she could no longer build and develop new ships at will, the agreement did at least ensure she remained the world's premier naval power and, for now at least, eliminated the prospect of another expensive naval arms race she could no longer afford.

Just because, for the first time in its history, the Royal Navy was restricted by an international agreement, the Senior Service still had a reputation to restore and set about doing so in the same year the Washington Treaty was signed.

Like HMS *Hood*, the Nelson Class was about restoring Britain's battered naval prestige, which had taken such a pounding in the First World War.

While they didn't exceed what was permitted by the treaty, they pushed the agreement to its limits and incorporated some novel design features to make sure they were still playing by the rules.

Equipped with nine 16-inch guns mounted in three triple-turrets, twelve 6-inch guns, anti-aircraft weapons and torpedo tubes, *Rodney* and her sister ship *Nelson* were, in the Royal Navy's glamour stakes, bridesmaids to the bride *Hood*.

The 16-inch guns, manufactured at Sir W. G. Armstrong Whitworth & Co. on the Tyne, Vickers of Barrow-in-Furness, William Beardmore & Company on the Clyde and the Royal Gun Factory at Woolwich, were formidable weapons that could fire their shells 788 metres per second up to a distance of nearly 40,000 yards and penetrate 14 inches of armour. It was reckoned the blast from the guns could break windows 5 miles away.

Unusually for British warships, all of the main guns on the Nelson Class were situated forward of the bridge to reduce armour weight and keep *Nelson* and *Rodney* on the right side of the treaty.

Another novel feature of the class was the large octagonal-shaped superstructure, rather like those found on modern aircraft carriers, that was referred to as 'Queen Anne's Mansions' due to its similarity to a large Victorian residential development called Queen Anne's Mansions near London's St James's Park. The superstructure housed the bridge, facilities for gunnery and torpedo control, accommodation and office space.

The design for the two ships was approved on 6 February 1922, and the invitation to tender for construction issued on 16 October. The Admiralty awarded the contract for building *Nelson* to Sir W. G. Armstrong Whitworth & Co. and *Rodney* to Cammell Laird at Birkenhead. During the early years of the twentieth century Cammell Laird had grown rapidly to become one of the UK's premier shipyards, alongside the fabled companies lining the banks of the Clyde, Tyne and Wear.

The two sisters were both laid down on the same day – 28 December 1922 – and thousands of workers set about building them into life, grateful for the work that was now far less frequent following the shipbuilding slump that came after the immediate post-war boom when lost merchant vessels had been rapidly replaced.

Rodney was launched on 17 December 1925 by Her Royal Highness Princess Mary in front of a crowd of some 20,000. During fitting-out every single part was closely scrutinised by the yard to save weight and remain within limits of the treaty. Such was the level of scrutiny the Nelsons became jokingly known as the 'cherry tree class' because 'they'd been cut down by Washington'.

Cut down they might have been, but *Rodney* and *Nelson* had still blossomed into magnificent vessels – both in appearance and the power they were capable of unleashing from their 16-inch guns.

An officer serving aboard *Rodney* was awestruck by his ship. He said:

> I challenge anyone who claims to possess what I am pleased to term a soul, to stand on *Rodney*'s fo'c'sle and contemplate the stark grey mass of turret and gun that stretches away before him – I challenge him to stand there by himself and not feel a definite tingle of pride and fear.

Midshipman Frank Rodham Twiss was also mightily impressed by her size and grandeur:

> The size of the gunroom and the ventilation of our chest flats and the general smartness of a new ship was the most prominent point I noticed. In fact the ship seemed three times as large as the *Revenge* … After supper in the gunroom I turned in, but the noise of electric motors and fans seemed at first so loud that I slept very little.

During trials, *Rodney*'s performance was mixed. Her guns more than matched expectations but her steering proved erratic and the ship difficult to control while steaming at slow speeds.

Rodney moved to her homeport of Plymouth in the summer of 1927 and was commissioned in the November of that year. On commissioning she joined the 2nd Battle Squadron of the Atlantic Fleet and took part in the regular deployments and cruises that made up the Atlantic Fleet's duties in the interwar years. Spring cruises tended to consist of exercises in the Mediterranean and Caribbean while summer cruises would frequently be spent in British waters and Scandinavia.

Like HMS *Hood*, crew on HMS *Rodney* took part in the Invergordon Mutiny in September 1931 and refused to put to sea over government plans to cut Royal Navy pay as the Great Depression bit through the bone and into the marrow of the nation's finances. As we've already seen in the chapter on *Hood*, the strike only ended when the Cabinet backed down and agreed not to reduce wages by as much as originally planned.

In July 1935 *Rodney* joined the navy's other premier ships in the Solent for King George V's Silver Jubilee Review and three years later she became the first British battleship to be fitted with radar. Radar – which uses radio waves to determine the speed, range, size and altitude of objects – was one of history's great inventions and discoveries. That *Rodney* was the first major vessel in the fleet fitted with this revolutionary piece of equipment was a measure of her status and importance to the Royal Navy. *Rodney* and her radar set were about to be tested like never before when the world went to war again for the second time in twenty-one years.

On 3 September 1939 the Home Fleet was in the North Atlantic 300 miles south of Iceland when the Admiralty signalled 'Total Germany'. Midshipman Charles Fetherston-Dilke was on *Rodney*'s bridge as the message came through:

> Total Germany was the code word for commence hostilities against Germany and there was comparative calm on the bridge, no one burst into tears or had to go to the loo … I suppose because it was expected … There was no gung-ho about it all just acceptance. That is what you are paid for … what you are here for. That is what the navy does when the shooting starts.

A chapel could be found on board *Rodney* – the Church of St Christopher, the patron saint of travellers. Given the 156,000 nautical miles she steamed around the globe during six long years of war, and the hits and near misses she survived, St Christopher more than repaid the honour.

From the start of hostilities, the main task of the Home Fleet, which was the new name for the Atlantic Fleet, was to prevent German ships and submarines breaking out into the Atlantic shipping lanes from where they could wreak destruction on the merchant supply ships Britain depended on for her survival. As in the First World War, the vast expanse of sea between Britain and Norway and Denmark and Iceland would be the route the Germans would try to pass through and the area the Royal Navy would have to defend from its main anchorage at Scapa Flow.

Rodney's first wartime action came on 26 September 1939, some 150 miles off the coast of Norway, in an action designed to lure the Kriegsmarine into a fight in open seas. Having learnt the lessons from the previous war, when the Admiralty had been content to hole the fleet up at Scapa and keep a watchful eye on the Germans rather than confronting them head on, this time the Senior Service was itching for a fight and determined to seize the early initiative.

Aircraft from the carrier *Ark Royal* were launched against German flying boats but the fleet's position had been identified and the Luftwaffe dispatched to attack. Five Junkers Ju 88 bombers swooped down on the powerful British armada that included *Hood*, *Nelson* and *Rodney* but all their ordnance splashed harmlessly into the water and they were driven off by the anti-aircraft barrage the ships put up.

Apart from the occasional minor action, the Phoney War was a time of tedium for the armed forces and many in the Home Fleet spent their days and nights as their fathers had done a quarter of a century earlier – bored and cut off from civilization at Scapa Flow. More used to the many and varied attractions on offer ashore at the likes of Plymouth, Portsmouth and Rosyth, winter in the Orkney Islands held little appeal and could be a difficult ordeal – even harder to cope with for some than the stresses and strains of battle.

Rodney's next notable encounter didn't take place until April 1940 during the German invasion of Norway. While steaming off the Norwegian coast near the city of Stavanger she was attacked by German aircraft and hit by a 500 kg bomb that pierced her armoured deck but miraculously failed to explode.

Following the fall of Norway and then the Low Countries and France in the summer of 1940, Britain braced itself for the impending Nazi attack and invasion of the British Isles. Key to the success of Operation Sea Lion, codename for the German invasion of Britain, would be getting thousands of troops safely across the English Channel and discharging them at landing points across the south coast.

To counter the anticipated invasion, *Rodney* was moved from Scapa to Rosyth, where she'd be within much easier reach of German shipping in the Channel. Although the RAF's victory in the Battle of Britain was seen as crucial to ending Germany's hopes of landing in England, even if the air war had been lost the invasion forces would still have had to contend with the Home Fleet and the destruction the 16-inch guns of *Nelson* and *Rodney* could have brought to bear. The might of the Royal Navy as well as the heroics of Fighter Command's young pilots undoubtedly played a role in preventing invasion so, even though she didn't fire a shot, *Rodney* and her fellow ships kept Nazi jackboots off British shores.

With the invasion threat lifted, *Rodney* returned to Scapa Flow and took on Atlantic escort duties, protecting convoys travelling between Halifax, Nova Scotia, and the UK at a time when the German U-boat wolf packs were claiming thousands of tons of Allied shipping every month.

To back up the work of their U-boats and menace the Allies from on, as well as beneath, the waves, the Kriegsmarine dispatched the battlecruisers *Scharnhorst* and *Gneisenau* towards the North Atlantic. Alerted by intelligence reports that the German surface raiders were on the move, the Home Fleet put to sea and lay in wait a hundred miles or so off the southern coast of Iceland. Whether the Germans knew where the British were and steamed on a different course to avoid them or the vastness of the cold, bleak north Atlantic meant they simply didn't detect each other is uncertain, but what is certain is the German ships broke through and sank dozens of British vessels over the next eight weeks as *Rodney* and the rest of the Home Fleet searched for them in vain. When *Scharnhorst* and *Gneisenau* were finally detected, *Rodney* and the brand new battleship *King George V* were sent to confront them but, hidden by bad weather, they escaped south-east and headed for the safety of Brest in occupied France.

If the first eighteen months of war had provided a mixture of boredom and frustration for *Rodney*'s crew, the next four years would prove far more dramatic and productive – starting with her key role in the destruction of Germany's newest, mightiest battleship.

In May 1941, *Rodney* was escorting the White Star Line's ocean liner MV *Britannic* to Canada and it was planned that, after delivering *Britannic* safely, she'd proceed to the United States for repairs and a refit. Yet while halfway across the Atlantic she received an urgent message from the Admiralty to join in the pursuit of the *Bismarck*, which had just sunk HMS *Hood* in the Denmark Strait.

After destroying *Hood* on 24 May and engaging in a firefight with *Prince of Wales*, *Bismarck* made for France knowing that every available unit of the British fleet would be in hot pursuit, seeking revenge for the destruction of the 'Pride of the Royal Navy'.

Spotted by a Coastal Command Catalina on 26 May, Swordfish aircraft from *Ark Royal* were launched and a single torpedo from one of the biplanes struck *Bismarck*, damaging her rudder and steering gear so badly that escape was now out of the question as she was unable to steer in a straight line.

With *Bismarck* floundering, *Rodney* and *King George V* moved in for the kill. 'We were glad to think, of all the British ships, that we would get the opportunity to avenge the *Hood*,' reflected a member of *Rodney*'s crew.

That revenge started at 8.47 a.m. on 27 May – the moment the massive guns of the British battleships opened fire.

'You didn't see nothing down there in the bottom of the ship. Our job was making sure that the weapons were lined up on target and ready to fire and then the director, he fired the guns. The ship was shaking every time she fired,' recalled teenage rating Mick Kavanagh, an Irishman who'd joined the Royal Navy in 1938. He went on:

> In my old Action Station, in the torpedo fire control position, I remember we used to have these rubber plugs that popped out of your ears when the big guns fired, so I suppose from that point of view I was better off down below. During the *Bismarck* action you had so much to do, so much to worry about getting right, you didn't think about what might be happening to endanger your life, nor did you reflect on what we were doing to the Germans.

As for the Germans, Burkard Baron von Mullenheim-Rechberg, a gunnery officer and the highest-ranking *Bismarck* survivor, gave an intense account of his ship's final moments in his memoir, *Battleship Bismarck*:

> The alarm bells were still ringing when, returning from the bridge, I entered my action station. I picked up the control telephone and heard, 'Two battleships

port bow.' I turned my director and saw two bulky silhouettes, unmistakably *King George V* and *Rodney*, at a range of approximately 24,000 metres. As imperturbable as though they were on their way to an execution, they were coming directly towards us in line abreast, a good way apart, their course straight as a die. The seconds ticked by. Tension and anticipation mounted. The nerves of our gun directors, gun captains, and range-finding personnel were steady. After the utterly hopeless night they had just spent, any action could only be a release. The very first salvo would bring it. How many ships were approaching no longer meant anything; we could be shot to pieces only once.

As if to compound the revenge, *Rodney* and *King George V* were joined by the heavy cruisers *Norfolk* and *Dorsetshire* to add their firepower to what was now one of the most one-sided battles in the history of naval warfare. A single 16-inch salvo from *Rodney* killed most of *Bismarck*'s senior officers by destroying the forward control post.

Len Nicholl, a Marine on board *Rodney*, witnessed the destruction:

I saw her battle flag flying when we first sighted her. Soon she was burning and I quite clearly saw one of the 16-inch shells hit and blow the back of a turret and it flipped around. It was such a rough sea that every time *Bismarck* was pushed under the water, flames would go out, but then she would come up and the fires would start again. I was on the port side and I think the *Rodney*'s 6-inch guns were doing more damage than our 16-inch shells which, being armour-piercing, were just going straight through *Bismarck*. We were so close, the joke later was that some of the lads on the upper deck could have thrown spuds at her.

By now it was clear to Mullenheim-Rechberg and the other survivors that their ship was doomed. He recalled:

I did not allow the men in my station to leave while shells were exploding all over the superstructure and main decks, and ready ammunition was blowing up … To do so would have been nothing less than suicide. I did not give the order to leave until long after our guns had fallen silent, the enemy stopped firing and, presumably, the shooting had come to an end. By this time our list to port was heavier than ever and starboard was the lee side. I called to the men to look for a place aft and to starboard on the main deck. Forward, there was too much destruction and the smoke was unbearable. The quarterdeck was out of the question: the sinking ship was too far down by her stern and heavy breakers were rolling over her from her port quarter. Those who made the mistake of jumping from that side or who were washed overboard in that direction were thrown back against the ship by the sea, in most cases with fatal consequences.

During the battle *Rodney*'s 16-inch guns fired 378 shells and she let off 706 6-inch projectiles, many of which were witnessed by Mullenheim-Rechberg. He wrote in his memoir:

> To continue, in the words of an observer in *Rodney*: 'From about 09.36 until cease firing at 10.16 *Rodney* steamed back and forth by the *Bismarck* at ranges between 2750 and about 4500 yards firing salvo after salvo of 16 inch and 6 inch during this entire period.'
>
> The trajectories of the shells were nearly flat and the devastation of the *Bismarck* was readily visible to her enemies. Several fires were raging and the back of turret Bruno was missing. The superstructure had been destroyed, men were running back and forth on deck, vainly seeking shelter, their only escape from the hail of fire being over the side.
>
> Around 10.00 the *Bismarck* appeared to the British to be a wreck. Her gun barrels pointed every which way into the sky, and the wind drew black smoke out of her interior. The glow of fires on her lower decks shone through the holes in her main deck and upper citadel armour belt.

After destroying the *Bismarck*, *Rodney* picked up from where she'd left off before being diverted and headed to Massachusetts for refit and repairs. On rejoining the fleet she was stationed in Gibraltar, escorting convoys to the besieged British colony of Malta.

As well as her 16-inch guns *Rodney* had a fearsome array of anti-aircraft weapons and she was used increasingly in an anti-aircraft role during her time in the Mediterranean.

Allan Snowden, a torpedo operator, made sure the anti-aircraft guns were provided with shells. As he recalled:

> The raids were quite exciting ... Usually they happened at night when I pulled duty as an ammunition supplier to one of the pom-pom close-range anti-aircraft weapons. You could see the German aircraft every now and then when they were caught in the searchlights, flashing in and out as they made their runs at us. I used to rush out with more ammo for the gun but was more at risk from our own shrapnel falling down than enemy bombs. It came pitter-patter down, caused by our ship's AA shells, or those of other vessels, exploding overhead. Strangely, I never heard of anyone being injured. You had tracers every fifth or sixth shell so it was quite a light show too and it was very noisy. We had received a lot of extra anti-aircraft weaponry aboard and the idea was that a ship put up a curtain of steel.

The amphibious landings the Allies conducted as the tide of war turned in their favour were perfectly suited to *Rodney*. Before troops went ashore in Operation

Torch in North Africa and then at Sicily and Salerno, her guns softened up enemy positions on and around the beach heads and rained down fire on Axis strong points and vehicles.

In June 1944 she took part in the biggest amphibious invasion of them all, sailing with the D-Day task force setting out to liberate occupied Europe.

Lieutenant-Commander Ian Hamilton was *Rodney*'s official battle observer for Operation Overlord, tasked with watching and recording the ship's course, orders given to the guns and the success or otherwise of the projectiles they fired. He was filled with both pride in his ship and the sense of occasion.

As *Rodney* neared Normandy he wrote:

> I am not anxious. I am not even nervous. On the contrary, I am elated. I am privileged. At this moment I would not change places with anyone. I am standing on the bridge of HMS *Rodney* as we steam towards the coast of Normandy … I see unfold before my eyes one of the great events in the history of our time.

Allan Snowden recalled the noise of the vast Allied armada opening up on the German defences along Hitler's Atlantic Wall:

> The sheer volume of noise … the blast of the guns, was incredible and you could feel it through your body even if you were quite a distance from the gun actually doing the firing. A bit nearer the coast were these rocket ships, which were like huge landing craft. When they fired it was like Guy Fawkes' night and you could see the rockets whooshing off into the air. You couldn't help feeling a bit sorry for the guys on the receiving end.

Coming under fire from a shore battery near the port of Le Havre, *Rodney*'s 16-inch guns replied in kind and the battery was no more. On D-Day plus one she took part in her first official bombardment of the Normandy campaign, shelling enemy positions a couple of miles north of Caen. Capturing the city was one of the first objectives of the British forces that had poured ashore from Gold, Sword and Juno beaches, so taking out German controlled points around it was crucial if Caen was to fall quickly and with as few casualties as possible.

Throughout June and early July, *Rodney* repeatedly moved back and forward from the gun line, heading back across the Channel to take on fresh supplies of ammunition before steaming to Normandy again to open fire. Her guns claimed numerous Panzers and had an incalculable role in destroying German morale. The 16-inch naval guns were the most powerful weapons in the Allies' armoury and repeatedly being on the receiving end of their fire had a profound psychological impact on the German defenders.

With Caen finally in Allied hands, and the British, American and Canadian armies breaking out of Normandy, *Rodney* returned home – but her next mission would see her firing at British territory. The Channel Islands were, and are, a UK Crown dependency but when France fell in 1940 they were captured by the Germans. Now it was time to take them back and *Rodney* was called in to take out four German guns on the island of Alderney.

As the war drew to its close the now battle-weary battleship escorted a convoy carrying supplies to the Soviet port of Murmansk and then returned to Scapa Flow, where she remained until after VE Day. Having steamed so many miles during the war without undergoing any extensive refits she had now reached the end of her useful life. She was decommissioned in 1946 then scrapped two years later.

If the *Bismarck* had survived after sinking *Hood* it might not have ultimately altered the outcome of the war, but it would have made *Hood*'s loss all the more sickening. What *Rodney* did was help cure the sickness and for that the nation was grateful.

HMS Rodney. Note the unusual shape of the superstructure.

This image captures *Rodney*'s size and shows how far back on the hull the superstructure sits.

Rodney entering Valletta harbour.

Rodney firing a broadside.

Rodney.

CHAPTER SEVEN

HMS *Illustrious*

HMS *Illustrious*
23,000 tons
Built by Vickers-Armstrongs, Barrow-in-Furness
Served 1940–1955

For centuries battleships ruled the waves as the ultimate symbols of naval power. From the Spanish Armada to the Battle of Jutland, the ships with the thickest hulls and biggest guns decided the fate of nations.

What the lion is to the jungle, so the battleship was to the sea. Then, flying through the moonlit skies of the Mediterranean in November 1940, twenty-one slow, flimsy biplanes swooped low on the Italian fleet anchored in Taranto harbour and claimed the battleship's crown for good.

This strike by Fleet Air Arm Fairey Swordfish was the first all aircraft ship-to-ship naval attack in history and showed that the future of naval warfare lay in the air above the seas and oceans instead of on the waves.

Said Admiral Andrew Cunning, Commander-in-Chief of the Mediterranean Fleet: 'Taranto, and the night of 11/12 November 1940, should be remembered forever as having shown once and for all that in the Fleet Air Arm, the Navy has its most devastating weapon.'

The aircraft that delivered the decisive blow had flown from the Royal Navy's new aircraft carrier HMS *Illustrious*. Her role in launching this game-changing attack is enough alone to secure her place in the annals of Royal Navy greatness. As it was, she went on to further justify her claim as one of the fleet's all time finest by surviving a ferocious bombing attack and serving in a variety of theatres throughout the Second World War.

The attack on Taranto wasn't the start of naval aviation, nor was *Illustrious* the first aircraft carrier. As early as the turn of the twentieth century, the Admiralty was investigating the possibility of launching aircraft in the form of manned kites from the decks of its warships to carry out spotting duties – essentially an early, primitive type of aerial reconnaissance.

Following the Wright Brothers' first successful powered flight in 1903, the world's great powers quickly realised the potential military value of aircraft for both offensive and defensive operations. In 1908, Britain's Prime Minister

Herbert Asquith approved the formation of an 'Aerial Sub-Committee' of the Committee of Imperial Defence to investigate the potential for naval aviation and look at how aircraft could be used at sea. A year after it was formed the sub-committee gave the go-ahead to the development of a rigid airship, resulting in the construction of *Mayfly*.

Appropriately, given the impact they both had on naval aviation, *Mayfly*, like *Illustrious*, was built by Vickers in Barrow-in-Furness on England's north-west coast. Unlike *Illustrious*, however, her career proved short-lived as the airship broke in two while being towed from her shed to conduct full trials. Despite never fully getting off the ground, *Mayfly* provided valuable experience and data that influenced later aircraft designs.

In 1910 Britain set up a naval flying school to train pilots and develop methods for using planes at sea. This led to the battleship HMS *Africa* becoming the first British ship to launch an aircraft when a Short Pusher took off from a downward sloping runway attached to her foredeck and flew out over the River Medway in Kent in January 1912. To modern eyes the Short Pusher looks like a bicycle attached to canvas wings, but at the time this was cutting-edge technology – and Britain was leading the way with it. HMS *Africa* had been stationary at the time of the launch, but in May 1912 the Royal Navy went one better and launched a Short Pusher from HMS *Hibernia* while she was steaming at 10.5 knots during the Royal Fleet Review.

While the trials had been successful, attaching a runway to a battleship's deck reduced the punch the ship packed as it made the forward guns unusable so alternative options were explored.

Seaplanes that could take off and land on water without the need for a ship-mounted flight deck were deemed a more suitable method for deploying aircraft at sea, so a seaplane base was established on the Isle of Grain in Kent followed by further bases along Britain's east coast at Calshot, Cromarty, Felixstowe and Great Yarmouth.

To extend the range of the seaplanes and enable them to operate on any sea or ocean, the next development was to convert the cruiser HMS *Hermes* into a seaplane carrier capable of accommodating three aircraft. As she was specifically designed to be an aircraft carrier, protected by other warships, it didn't matter that her forward gun would be unusable so the armament was removed.

The successful conversion of *Hermes* to the seaplane carrier role led to the Admiralty then ordering a purpose-built carrier. HMS *Ark Royal*, a name that became synonymous with British naval aviation, was initially laid down as a merchant ship but was purchased by the Royal Navy shortly after work on the keel got underway. This allowed the vessel's design to be radically altered and modified specifically for carrying seaplanes. It made *Ark Royal* the first purpose-built aircraft carrier in the world and highlighted yet again Britain's dominance over every other power in sea warfare. Commissioned in December

1914 she served throughout the First World War, supporting Allied forces at Gallipoli and providing air cover for the British Army on the Macedonian front.

While *Ark Royal* proved a useful platform for deploying and recovering seaplanes, plans and designs for operating aircraft with wheels from either adapted ships or new aircraft carriers continued. The Courageous Class battlecruiser HMS *Furious* was modified while under construction and launched in August 1916. Extensive trials using Sopworth Pup planes led to her introduction into service in mid-1918. During trials Squadron Commander Edwin Harris Dunning became the first man to land a plane on a moving ship when he touched down on *Furious* at Scapa Flow on 2 August 1917.

At the same time *Furious* was under development, the Admiralty was also working on a design that's broadly similar to those of aircraft carriers in service today. HMS *Argus* started life at William Beardmore's shipyard on the Clyde as the Italian ocean liner *Conte Rosso* but when building work was halted at the start of the war the Royal Navy acquired the hull and the yard built her as an aircraft carrier with a full-length flight deck that could carry between fifteen and eighteen planes. *Argus* was commissioned in September 1918 and served on-and-off until 1946.

In the early 1920s HMS *Glorious* and HMS *Courageous*, which like *Furious* had started out as Courageous Class battlecruisers, were converted into aircraft carriers and HMS *Eagle* was commissioned in 1924, giving the Royal Navy five aircraft carriers – the largest carrier fleet in the world. By the 1930s though, with more powerful planes coming into service and Germany re-arming after Hitler swept to power, the navy identified a need for a new aircraft carrier that could carry a far greater number of planes.

On 16 September 1935 the keel of HMS *Ark Royal* (the original First World War era carrier had been renamed *Pegasus* to free up the *Ark Royal* name for the new ship) was laid at Cammell Laird's shipyard at Birkenhead on the River Mersey. Unlike the earlier carriers, *Ark Royal*'s hangars and flight deck were built as part of the vessel's superstructure rather than being added on to the hull and she was fitted with steam catapults and arrestor wires for launching and landing her complement of fifty aircraft.

Having set the benchmark for the modernised aircraft carrier design, and with Britain finally recognising that it needed to speed up the expansion of its armed forces to counteract the growing Nazi menace, *Ark Royal* was followed down the slipways by a new class of four more carriers – *Illustrious*, *Formidable*, *Victorious* and *Indomitable*, all of which were laid down in 1937.

While *Illustrious* and her sisters bore many similarities to *Ark Royal*, they had the additional protection of an armoured flight deck and hangars that *Ark Royal* lacked. As we shall see, this armour proved crucial.

The contract to build the lead ship of the class, *Illustrious*, was awarded to Vickers-Armstrongs at Barrow-in-Furness. Launched on 5 April 1939 and christened by Lady Henderson, wife of the recently deceased controller of the Navy who'd championed the cause of new aircraft carriers, she was then the biggest ship ever built in Barrow. Following her launch, tugs towed the hull to the yard's Buccleuch Dock where fitting-out work was speeded up to get her rapidly into service following the outbreak of war in September 1939.

Just four months after commissioning in April 1940, *Illustrious* left Scapa Flow bound for the Mediterranean with 806 Naval Air Squadron and its Fairey Fulmar fighters and the Fairey Swordfish-equipped 815 and 819 Squadrons embarked.

Malta was home to the Royal Navy's Mediterranean Fleet and of vital strategic importance to the British Empire due to its proximity to the Suez Canal. When Italy joined the war alongside Germany and Japan, Britain feared an Italian invasion of the colony and determined that it be fortified and defended. Protecting Malta shaped British policy in prosecuting the war in the Mediterranean and ultimately led to the Battle of Taranto.

On arrival in the Mediterranean, *Illustrious* participated in Operation Hats, escorting convoys from Gibraltar and Alexandria taking supplies to Malta. When the convoy came under attack, *Illustrious*'s Fulmars shot down five Italian aircraft and her anti-aircraft guns claimed two more. Then on 3 September, a year to the day since Britain and France had declared war on Germany, Swordfish from *Illustrious* and *Eagle* struck at an Italian seaplane base on the Greek island of Rhodes. The raid saw further success for the Fulmars, which were proving to be highly capable naval fighters, when they took down three of the Italian defenders.

Those onboard *Illustrious* could have been forgiven for thinking this war fighting was all rather easy, for they notched up another impressive feat when fifteen Swordfish attacked Italian forces at Benghazi harbour in North Africa. The first wave of aircraft attacked a destroyer and two merchant ships by dive-bombing, while the second group laid a minefield that claimed another destroyer, sank two merchantmen and badly damaged a third.

While the Fairey Fulmar looked every bit the modern fighter aircraft in 1940 – a monoplane with sleek lines that was capable of flying at nearly 300 mph – the Fairey Swordfish more closely resembled Orville and Wilbur Wright's machine that took to the air in 1903. Made out of a metal airframe covered in fabric, the open cockpit Swordfish was a biplane with a top speed of just 143 mph and primitive instruments little better than a compass, altimeter and air speed indicator. Capable of operating in a torpedo-dropping role, mine laying or as a dive-bomber, the planes had space for a crew of three – a pilot, observer and a radio operator/rear gunner – but usually only carried two.

Nicknamed the 'Stringbag' due to the bracing wires between the wings and its ability to carry an assortment of pieces of equipment – like a shopping bag – the Swordfish was introduced into service in 1936 and nearly 2,400 had been built by the time it was retired in 1945. It might have been a slow and plodding aircraft that looked like it hailed from another era, but its reliability, endurance and effectiveness at delivering torpedoes and bombs meant it was much loved by the crews that flew her.

Lord Kilbracken, who, as John Godley, flew Swordfish for four years during the Second World War, had very fond memories of the Stringbag and wondered at its unlikely capabilities. He recalled:

> A makeylearn pilot converting to any other operational type at once knew he was handling a lethal, steely machine that meant business ... Not so the Stringbag, bless her dear old heart. She seemed to have been left in the war by mistake, to belong to another age, though she had been in production for only six years ... Yet she had the most wondrous virtues. Was almost totally foolproof. You could take her nearly any liberty, fly her far beyond textbook capability. The willingness and ability to do so was part of being a worthy Stringbag pilot and she always saw you through. She was absolutely stable, and even at almost the lowest speeds the controls were firm and positive. Scream down from an immense height in a dive, the speed would stay well below 200 knots and you could haul back on the stick for all you were worth: a fast, firm pull-out with no fear of that old enemy the high-speed stall.

In some ways the Swordfish's slow speed worked to its advantage. Unlike other faster aircraft it could take-off and land in bad weather with there being less chance of the pilot losing control. Swordfishes were also highly durable and could continue flying even after being shot-up and hit by anti-aircraft fire as the enemy bullets and shells would often pass straight through the fabric wings and fuselage without destroying the plane.

The performance of the Swordfish aircrews at Taranto and number of hits they recorded on the Italian fleet made it seem like dropping torpedoes was as straightforward as shooting fish in a barrel. In reality it was an art and a skill that could only be acquired through many hours of practice. Godley went on:

> The trickiness lies in the fact that you have to be flying in the correct low speed, and as straight and level as possible, and at the right altitude, at the moment you release it and for several subsequent seconds, if it's going to run accurately or indeed run at all.
>
> Another little matter is that over a minute will elapse if you drop from as close as 1,000 yards, before the tinfish reaches its target which will meanwhile have travelled at 800 yards if streaming at 20 knots. So you must judge her speed, aim the right distance ahead and guess what avoiding action she might take.

Practicing on torpedo and bombing ranges might teach and then hone the skills required but all the training in the world can never replicate what those going to war face. The time had come for Admiral Cunningham to test his pilots in combat.

Cunningham knew that Italy's naval base at Taranto posed a major strategic threat to British forces in north Africa and the Mediterranean. By the summer of 1940, the Italian Navy had two 13,000-ton and four 14,000-ton cruisers with 8-inch guns and six battleships. It was these battleships the British most feared. Two of the battleships – *Littorio* and *Vittorio Veneto* – were brand new and boasted a fearsome array of firepower. Fully loaded, they each displaced 45,000 tons and had a crew of 1,800. Each ship had triple turrets, each mounted with three 15-inch guns.

In theory, Cunningham's plan for dealing with this threat was simple. Taranto harbour lies in the great bight of the Gulf of Taranto – a position from where any ship approaching would be detected far out to sea and could easily be repulsed. With the harbour being too shallow for submarines and unsuitable to attack with warships, an air strike was Cunningham's only option if the Royal Navy was to launch a surprise strike when the Italian fleet was at anchor.

In the months leading up to the night of the attack, Britain had endured a torrid time. France and the Low Countries had all fallen that summer, leaving the British Army cut off and in need of rescue from the beaches of Dunkirk. Denmark and Norway had also come under Nazi control with British forces taking another heavy beating at German hands in Norway. A military victory was required to give the nation, becoming used to defeat and retreat, a morale boost, so the pressure was on the Royal Navy to deliver at Taranto.

Initially it was planned for the ageing First World War-era carrier HMS *Eagle* to join *Illustrious* in launching aircraft against the Italians but a breakdown in her fuel system took her out of action and left *Illustrious* to go it alone. Five Swordfish from *Eagle* were transferred to *Illustrious* before the battle to increase the number of aircraft available.

The complete task force, commanded by Rear-Admiral Lumley Lyster, for Operation Judgement – as the Taranto attack was codenamed – was made up of the heavy cruisers *Berwick* and *York*, the light cruisers *Gloucester* and *Glasgow* and destroyers *Hyperion*, *Ilex*, *Hasty* and *Havelock*. Confirmation of the Italian fleet's presence in Taranto was confirmed by a reconnaissance plane on 10 November, so the task force raised steam and headed for the launch point some 170 miles south-east of the harbour.

Under fine, moonlit conditions, the first strike of twelve Swordfish took off – six armed with torpedoes and the others carrying bombs. The plan was to hit the battleships in the outer harbour with torpedoes and use bombs on smaller vessels in the shallower waters of the inner harbour.

It had been a calm and peaceful night for those on board the Italian ships until flares illuminated the skies over them not long after 11 p.m. Then, swooping out of the moonlight and plunging down to thirty feet above the waves, came the Stringbags. A wall of anti-aircraft fire went up from the ships and shore gun emplacements below but it lacked accuracy and the attackers got through to discharge their payload of torpedoes and bombs.

A second wave of nine Swordfish had been dispatched from *Illustrious* at 9.23 p.m. and as they approached Taranto they saw the reassuring sight of a blazing inferno inflicted by the first attack. Again the Italian guns opened up and again they were largely ineffective as the aircraft swooped and swerved their way through to hit ships in the harbour and oil tanks ashore.

Lieutenant Michael Torrens-Spence was a pilot aboard *Illustrious* and described the experience of flying in the raid at Taranto in the book *Lost Voices of the Royal Navy*. He wrote:

> Lieutenant-Commander Williamson led off the first wave of aircraft. My squadron commander, Ginger Hale, led the second wave about half an hour later.
>
> During the attack a hundred thousand rounds were fired at us but only one aircraft was shot down in each wave. I got hit underneath by one half-inch machine-gun bullet. It was the pilot's job to aim the torpedo. Nobody was given a specific target. I dived down in between moored ships aimed at the nearest big one, which turned out to be the *Littorio*, and released my torpedo. While you're low down over the water and surrounded by enemy ships the comfort is that they can't shoot at you without shooting at each other. I then made for the entrance to the harbour at zero feet and thence back to the *Illustrious*. It's difficult to know whether you have hit the target or not because, once you have dropped the torpedo, you're away. I didn't even see any of the action around me, I was too busy looking for barrage balloons.
>
> There was great jubilation aboard *Illustrious* at having had only two aircraft shot down. When we were debriefed they wanted to know if I had hit anything and I could only say that you couldn't miss a big ship at such a short range if the torpedo ran straight. I was credited with a hit on the *Littorio*.

As night turned to day the Italians were left to come to terms with the mauling of their fleet. *Littorio*, struck by three torpedoes, was listing heavily and so badly damaged she remained out of action for over a year. The battleship *Caio Duilio* had been holed and only prevented from sinking by the prompt actions of her crew beaching her while the *Conte de Cavour* sank to the bottom of the harbour. Although eventually raised, she'd take no further part in the war. In addition to the big ships, the heavy cruiser *Trento* and a number of destroyers had been smashed by bombs.

The British were elated, the Italians deflated.

Italian Foreign Minister (and Mussolini's son-in-law) Count Ciano wrote in his diary:

A black day. The British, without warning, have attacked the Italian fleet at anchor in Taranto, and have sunk the dreadnought *Cavour* and seriously damaged the battleships *Littorio* and *Duilio*. These ships will remain out of the fight for many months. I thought I would find the Duce downhearted. Instead he took the blows quite well and does not, at this moment, seemed to have fully realised its gravity.

In the House of Commons, Churchill, with as good a sense for the implications of an action on future events as ever, recognised the gravity of what had been achieved and the impact it would have on naval warfare. He commented:

As the result of a determined and highly successful attack, which reflects the greatest honour on the Fleet Air Arm, only three Italian battleships now remain effective … This result, while it affects decisively the balance of naval power in the Mediterranean, also carries with it reactions upon the naval situation in every quarter of the globe.

Never before had warplanes flying from an aircraft carrier delivered such a devastating blow to an enemy fleet. Its impact was so great that it would be copied on a larger scale thirteen months later in an operation that altered the entire course of the war. Thousands of miles away in the Far East, commanders of the Japanese Imperial Navy poured over details of the successful attack. Lieutenant-Commander Takeshi Naiton, Japan's assistant naval attaché to Berlin, flew to Taranto to investigate the attack first hand and subsequently had a lengthy conversation with Commander Mitsuo Fuchida about his observations. In December 1941 Fuchida would lead Japan's attack on the United States Pacific Fleet at anchor in Pearl Harbour.

Illustrious and her Stringbags inflicted such damage to the battle fleet that it was a defeat from which Italy never fully recovered. Before Taranto, Mussolini harboured ambitions of taking on and beating Britain to leave the Mediterranean at his mercy. The loss of his battleships, though, meant the Royal Navy could operate in the region at will.

Admiral Cunningham was certain Taranto had been a game-changing moment. His dispatch to the Admiralty in January 1941 stated:

There can be little doubt that the crippling of half the Italian battle fleet is having, and will continue to have, a marked effect on the course of the war. Without indulging in speculation as to the political repercussions, it is already evident that this successful attack has greatly increased our freedom of movement in the Mediterranean and has thus strengthened our control over the central area of this sea. It has enabled two battleships to be released for operations elsewhere, while the effect on the morale of the Italians must be considerable. As an example of economy of force it is probably unsurpassed.

With Italy now unable to take on the British alone following Taranto, and the capitulation of Mussolini's army in North Africa to Britain's small but well-trained Western Desert Force, the Nazis came to the rescue of their Italian allies.

Hitler dispatched his Afrika Korps, under the command of General Erwin Rommel, to North Africa and bolstered Axis forces in the Mediterranean.

Her success at Taranto had made *Illustrious* a prized scalp for the increased German presence in the region and in January 1941 the Luftwaffe found her and struck. Lieutenant Torrens-Spence had this to say:

> We knew that the Germans had sent a Fliegerkorps of Stukas to Sicily, commanded by a Luftwaffe Major-General … Their aim was to sink the *Illustrious*, having appreciated that she was the key to the control of the Mediterranean, but by now we were down to twelve Fulmars, having started with eighteen, and of these only five were serviceable.
>
> I landed one at lunchtime after an anti-submarine patrol and went down to the wardroom for lunch. About ten minutes later a wave of Stukas arrived and a 1,000 lb bomb penetrated the armoured deck and went off in the hangar above the wardroom. In the consequent smoke, black-out, and with red-hot shrapnel flying about, I made a dive for the wardroom door and got up on deck. It transpired later that half a dozen people who were having lunch with me were killed. When the attack developed we had two Fulmars on patrol, but they had been brought down to chase some Italian torpedo-bombers, and three Fulmars on deck, which took off only seconds before the first bomb hit, so the Stuka attack was virtually unopposed. The Fulmars laid into the Stukas and shot down a few, but it was nearly a miracle that the ship got back to Malta, having taken five bombs and having the steering gear jammed.

Of the five bomb hits, the strike in the hangar above the wardroom where Torrens-Spence was eating his lunch was the most damaging. The explosion started a ferocious fire and blew such a large hole in the hangar deck that parts of the ship three decks below were damaged. Had *Illustrious*, like *Ark Royal*, lacked an armoured deck she'd almost certainly have floundered.

Believing they'd destroyed her, the German planes called off the attack and returned to Sicily while *Illustrious* limped back to Malta. Torrens-Spence went on:

> The Germans thought they had sunk us, which is partly what saved us … The Luftwaffe pilots went back and said 'We have sunk her,' so they stopped sending in more attacks. Funnily enough when we were in Greece a few months later, I spoke to a German Stuka pilot, a prisoner of war, who had been bombing us a few months earlier. Even then he would not believe that the *Illustrious* had not been sunk.

While the Luftwaffe had failed to deliver a mortal blow, the damage was extensive. Emergency repairs were carried out at Valetta harbour in Malta and she then crossed the Mediterranean to Egypt for a further temporary fix at Alexandria. The patch-up completed, *Illustrious* headed for the United States via the Suez Canal and around the African cape for extensive repairs and maintenance in Norfolk, Virginia, that lasted six months.

Once the overhaul was complete, *Illustrious* departed for the UK in December 1941, sailing alongside her sister carrier HMS *Formidable*. Like *Illustrious*, *Formidable* had been repaired in America having been attacked by Stukas in the Mediterranean. During the journey back across the Atlantic the sisters collided in a storm and, although neither ship was badly damaged, *Illustrious* needed repairing yet again – this time at Cammell Laird's shipyard in Birkenhead.

Finally back in service in March 1942, she was sent to the Indian Ocean to assist in the Allied invasion of Vichy French-held Madagascar. The invasion was launched due to fears that the Vichy French would allow Japanese forces the use of the island and *Illustrious*'s aircraft were tasked with attacking enemy naval units and providing air cover to the Allied fleet.

From Madagascar she was then formally assigned to the Eastern Fleet for six months before returning to Britain in January 1943. In August 1943 *Illustrious* went back to the Mediterranean for the first time in more than two years, protecting the American, British and Commonwealth landings at Salerno on mainland Italy. With the Italians on the verge of defeat and switching sides, she didn't come under attack and the most noteworthy action her aircraft saw was escorting a surrendered Italian plane back to Sicily.

By late 1943, with the Red Army driving Hitler's forces out of the Soviet Union and the British and Americans advancing through Italy, such good progress was being made in the European war that Allied resources, including *Illustrious*, were freed up for action against Japan.

Departing the UK on 30 December, she arrived in Ceylon at the end of January 1944 and participated in several hunting trips with the Eastern Fleet, searching for Japanese warships around the Bay of Bengal. She then teamed up with the US Navy to launch operations in the Dutch East Indies (now Indonesia) and Andaman Islands. Instead of Fulmars and Swordfish, *Illustrious* now carried more modern and powerful Barracuda dive-bombers and Corsair fighters that she successfully launched against Japan's small naval base at Subang on the northern tip of the Sumatra on two separate occasions.

Following a brief refit in South Africa she was next assigned to the newly formed British Pacific Fleet and carried out further raids on Japanese forces in Sumatra. During one of the attacks, twelve men on board *Illustrious* were killed and twenty-one wounded when she was hit by friendly fire from either her fellow carrier *Indomitable* or the battleship *King George V*.

Repairs to the damage sustained in the friendly-fire incident were carried out in Australia to enable her to be back in service in time to support the American invasion of the Japanese home island of Okinawa.

By this stage of the war the Allies were able to deploy overwhelming force so, as well as *Illustrious*, the British Pacific Fleet made three other carriers – *Indomitable*, *Victorious* and *Indefatigable* – available. The Royal Navy's role in the operation was to neutralise Japanese airfields and wipe out aircraft on the ground while American troops landed on Okinawa.

As a last throw of the dice, Japan began launching kamikaze attacks against Allied ships. On the afternoon of 6 April 1945 a kamikaze pilot dived down on *Illustrious* but her anti-aircraft fire took the plane out seconds before it crashed onto the flight deck. However, the bomb carried by the kamikaze exploded in the water just yards from the ship and damaged her hull. The extent of repairs required meant she had to head back to Britain and the war ended shortly after she got home.

With peace restored and austerity looming, the Royal Navy had far more ships than it needed in a new post-war world where the loss of the British Empire and the United States and Soviet Union's rise to superpower status meant Britain's role had diminished.

Instead of sending her back to sea on routine deployments, the Admiralty designated *Illustrious* a training and trials carrier for the training of Fleet Air Arm pilots and testing new aircraft as they came into service. She continued in this role until being paid off at the end of February 1955.

Before Taranto, battleships dominated every major naval engagement. After Taranto, navies stopped building battleships and centred fleets around aircraft carriers instead. Not only did *Illustrious* and her Stringbags secure Britain a much-needed victory in the Mediterranean, they changed naval warfare and shaped the makeup of navies we see around the world today.

Fairey Swordfish landing on *Illustrious.*

A view of *Illustrious*'s stern.

Illustrious steaming at speed.

Swordfish taking-off and landing. (Credit Michael Pocock and MaritimeQuest.com)

CHAPTER EIGHT

HMS *Amethyst*

HMS *Amethyst*
1,350 tons
Built by Alexander Stephen & Sons, Govan
Served 1943–1952

HMS *Amethyst*'s daring dash to freedom down the Yangtze River made newspaper headlines around the world and turned her into a star of the silver screen.

She might have been a small, unglamorous vessel lacking film star looks, but the heroics of those on board *Amethyst* while trapped and targeted in a war zone were more than worthy of any movie idol.

Having found herself thrust from obscurity into the limelight when caught in the middle of the battle raging for control of China between the communists under Mao Zedong and the Nationalist government of Chiang Kai-Shek, she displayed qualities so admired by the British – doggedness, determination, resilience and winning through against all odds.

Being centre-stage in the global news was a world away from where *Amethyst* had started life on a Govan slipway.

Throughout the Second World War, Atlantic convoys carrying food, materials and munitions from North America to the UK were Britain's lifeline. As an island nation the British depended on imported goods and needed more than a million tons of material from overseas every week just to be able to survive and continue fighting.

Hitler knew only too well that Britain's reliance on imports was her Achilles heel so, from the moment war was declared, he despatched his U-boats into the Atlantic to menace the merchant ships bringing their vital supplies to British shores.

Hunting in groups, or wolf packs as they were called, German submarines would spread out in a long line across the predicted course of a convoy and then attack en mass when a target was sighted with the aim of overwhelming any escorting warships and picking off the merchant vessels.

Submarines of this era were slow under water but, on the surface, faster than any convoy of merchant ships, and their low silhouettes made them difficult to spot – especially at night or in bad weather.

Winston Churchill recognised early on that the whole course of the war would be determined by what happened in the icy waters of the North Atlantic. 'The Battle of the Atlantic was the dominating factor all through the war,' he wrote. 'Never for one moment could we forget that everything happening elsewhere, on land, at sea or in the air depended ultimately on its outcome.'

Allied shipping losses in the Atlantic reached a peak in 1942. In the first half of that year alone 1,000 vessels were sunk – an unsustainable rate if allowed to continue.

While the large battleships and cruisers frequently deployed to protect the convoys were effective in deterring surface raiders, they were less successful in dealing with the U-boat threat. Smaller vessels with more specialised anti-submarine capabilities were of much greater use. It was in boosting the numbers of these types of ships that HMS *Amethyst* came into service.

Laid down on 25 March 1942 at Alexander Stephen & Sons yard in Govan on the Clyde, *Amethyst* was a modified Black Swan Class sloop. Like corvettes, sloops were specialised convoy-defence ships, although sloops were larger, faster and equipped with superior anti-aircraft guns and a wider array of depth charges for destroying U-boats.

Weighing in at just 1,350 tons and powered by two Parsons turbines that could push her up to a speed of just over 18 knots, *Amethyst* had a crew of 190 and a range of around 5,500 nautical miles.

By the time she was commissioned in November 1943, the Battle of the Atlantic, and the whole course of the war, had begun to turn in the Allies' favour – and it was the U-boats that were being sunk in greater numbers than the merchantmen.

The major breakthrough had been technological advances in Allied equipment and weapons. Detection devices that could pick up and pinpoint where U-boat radio signals came from enabled better targeting of submarines and the introduction of shortwave radar meant U-boats could be detected on the surface. Improvements to Asdic sets, now known as sonar, which discovered enemy vessels under water by emitting pulses of sound and listening for echoes, allowed the Allies to track U-boats as they twisted and turned to take evasive action. The U-boats' slow underwater speed then made them highly vulnerable once they'd been traced.

New and improved anti-submarine weapons were coming into service, including the Hedgehog system developed by the Royal Navy. Hedgehog fired forward-throwing mortar bombs into the water that would explode on contact with a submarine. It proved a highly potent weapon, achieving more U-boat sinkings than the depth charge.

More effective use of air power – both Fleet Air Arm planes operating from carriers and RAF Coastal Command's long-range aircraft flying from land bases – against U-boats from 1943 onwards also played a key role in keeping the supply lines open and winning the Battle of the Atlantic.

German losses became so high that eight out of every ten Kriegsmarine submariners would die in action and U-boats became known as iron coffins.

While the war at sea was being won when *Amethyst* came into service, the fighting was still fierce and she was deployed regularly on convoy and anti-submarine duties.

Her main armament was made up of six 4-inch dual-purpose guns that could engage both surface and aerial targets mounted in pairs behind the partial protection of armoured shields. She also carried 20-mm cannons, Hedgehog and depth charges.

Amethyst's regular convoy route was in the Mediterranean, escorting supplies being transported between Port Said in Egypt and Gibraltar. After a brief refit at Devonport Dockyard lasting from November 1944 until early January 1945 she was assigned to the 22nd Escort Group, operating from Liverpool.

While escorting a convoy off the south-west coast of Ireland in February 1945 she attacked and sank the German submarine *U-1276*, then, the day after VE Day, she accepted the surrender of a German U-boat off Portland in the English Channel.

With war in Europe over *Amethyst* was deployed to the British Pacific Fleet in the Far East, where the battle with Japan was still raging. She operated alongside the US Navy and was present for the surrender of Japanese forces in New Guinea.

At the end of the war the Royal Navy was left with a vast number of ships, many of them no longer needed. Those surplus to requirements were either sold off to foreign navies or sent to the scrapyard. Being a more modern vessel that had only been in service for two years, however, meant *Amethyst* survived the cull. As ships weren't now required for convoy protection duties some of the sloops retained in the fleet, including *Amethyst*, were modified and redesignated as frigates.

As a frigate, *Amethyst* was assigned to the Far East Fleet. The fleet, based at Singapore, was responsible for protecting British interests and possessions across east and south-east Asia.

One of the tasks of the Far East Fleet was to provide the recognised guard ship for the British Embassy at Nanking, the seat of the Chinese government, some 200 miles up the Yangtze from Shanghai. It was while travelling up the great river to take over guard duties from HMS *Consort* that *Amethyst* found herself caught up in a war she had nothing to do with and thrust into the global spotlight.

The Chinese Civil War had first begun as far back as 1927 when fighting erupted between forces loyal to the Nationalist government of China and the Chinese Communist Party. In 1922 the foundation of the Soviet Union – which had been forged in the wake of the 1917 Russian Revolution as the world's first Marxist–Leninist state – inspired communist movements across the globe. A communist revolt led by Mao Zedong against the Chinese government began in 1927 following the formation of a communist rebel army, which later became the People's Liberation Army.

Over the next ten years the Nationalists and communists battled for control of their country, but in 1937 ceased hostilities to join forces against the invading Japanese army. The Sino–Japanese war, which effectively became part of the Second World War in 1941 after the Japanese attack on Pearl Harbour brought America into the conflict, claimed more than 20 million Chinese lives – including millions of civilians murdered in mass atrocities by the Japanese invaders.

With the defeat of Japan in 1945 the Nationalists and communists, despite all the blood that had already been spilled, picked up where they'd left off eight years earlier and restarted the civil war which was now swinging in favour of Mao's men thanks to the Soviet supply of equipment and weapons.

In 1949, when *Amethyst* arrived on station, the conflict was reaching its climax and would end a year later with mainland China falling fully under communist control and being renamed the People's Republic of China and the Nationalists fleeing to the island of Taiwan and relocating their seat of government to the Taiwanese capital, Taipei.

As *Amethyst* steamed up the Yangtze, the vast People's Liberation Army, spread out over a front of several hundred miles, stood on the north side of the great river. On the opposite bank the now tiring nationalist forces were lined up and waiting for the communist onslaught they knew was coming if they didn't give in to Mao's ultimatum and hand over their leader Chiang Kai-shek.

Having left Hong Kong on 19 April, the ship took on two Chinese river pilots. With its shifting sandbanks, numerous small islands and seasonal flood-waters the mighty Yangtze is a difficult waterway to navigate and, without local knowledge of its channels, currents and conditions, a ship can quickly run into difficulty.

Amethyst was commanded by Lieutenant-Commander Bernard Skinner, a thirtysomething who'd joined the Senior Service in 1937 after serving in the Merchant Navy with the Blue Funnel Line. He'd never see the wife he'd just said goodbye to again.

Shortly after sailing Skinner received a signal from the British naval attaché in Nanking informing him the Nationalists had refused Mao's surrender terms and that a communist attack across the Yangtze was therefore likely on the morning of 21 April. This, Skinner reckoned, would give *Amethyst* ample time to get safely to Nanking and for *Consort* to steam out of the way in the opposite direction.

At 5 p.m. *Amethyst* reached Kiang Yin, the point at which the estuary suddenly narrows to the waters of the river proper. As navigation of the Yangtze was even more hazardous by night and forbidden by the Chinese Navy, Skinner anchored his ship and waited until daybreak to get underway again.

Nearing the troubled area where the rival armies were lined up facing each other across the waterway, at around 8 a.m., Skinner gave orders to unfurl large canvas Union Jacks over both sides of the ship to make clear in no uncertain terms that this was a British vessel and nothing to do with the Chinese Civil War.

Speed was increased to 16 knots and crew called to action stations, so the ship was on high alert and steaming hard to get through the danger zone as quickly as possible.

Half an hour later the crackle of rifle fire could be heard and bullets splashed into the water near *Amethyst*'s hull. Then shells began landing around her and Skinner instructed his gunners to fire back in the direction of the attack. A dense mist had descended on the Yangtze though, and the British guns couldn't get a fix so stayed silent. Just as suddenly as the shooting and shelling had started, it stopped. None of the shells had hit her so *Amethyst* could breath easy.

When the firing ceased First Lieutenant Geoffrey Weston turned to his skipper and said: 'Don't think I need worry sir; I call that a first-class example of damned bad shooting.'

'Yes,' replied Skinner. 'You'd think they couldn't possibly have missed. If their other batteries are no better than that, we can all be easy.'

Further down the river, some forty minutes later, four shells struck the ship in quick succession. On the bridge Skinner gave orders for a hard turn to starboard and to 'open fire'. At the moment the words 'open fire' left his mouth the bridge was struck with two direct hits that killed or severely wounded everyone standing there. Skinner fell with catastrophic wounds to his shoulder, side and back. He died two days later.

First Lieutenant Weston had been on the bridge at the moment the shells struck and was hit by a splinter that ripped through his lungs and lodged in his liver. Despite the wound and coughing up blood, he remained conscious and displayed the type of courage and leadership synonymous with the Royal Navy since the days of Nelson, instinctively taking control.

Surrounded by dead, dying and injured comrades and with fire still raining down, he saw the ship lurching towards Rose Island, one of the many islets in the waters of the Yangtze. He ordered 'full astern both engines' but it was too late. *Amethyst* continued towards the island and her bows sunk into a mud-bank, leaving her a sitting duck for the communist guns.

As she came to a halt, Weston sent out the signal: 'Under heavy fire. Am aground in approx. position 31.10' North 119.50' East. Large number of casualties.' The position flashed was many miles from where the ship was lodged in the mud but the navigation chart had been torn to pieces and shells were continuing to land so it was hardly surprising the signal was inaccurate.

Shells exploded in the sick bay, port engine room and generator, causing further casualties. Few onboard expected anything other than *Amethyst*'s complete destruction. Due to the damage and angle she came to rest at, only one gun was able to return fire at the shore batteries blasting away at will.

'It seemed certain to me,' said Weston, 'that the communists were bent on bombarding the ship to destruction. There was now no action we could take to prevent it and every minute someone was being killed or wounded.'

Able Seaman Ray Calcott said:

Immediately when we came under fire, the skipper ordered the lashings to be cut and the massive Union Jacks were unfurled in order that no one could mistake which country we were from. At this time I was on the quarterdeck and dashed up to my action station, which was A-gun. At this point the shore battery kept on firing, hitting us at will. A and B-guns could not bear and only X-gun was able to fire, but it was soon put out of action. When the wheelhouse took a direct hit the wheel went over to port, the telegraphs were stuck and we went aground on Rose Island. Movement was restricted on the upper deck as snipers on the shore and their main armament continued to fire on us.

Sometime between 10 a.m. and 10.30 a.m. Weston ordered the immediate evacuation of anyone who could be spared. Some men swam while a boat ferried others including some of the wounded away from *Amethyst*.

Spotting the evacuation underway, Mao's marksmen turned their attentions from the ship to those heading for the shore, so the getaway had to be stopped. Around sixty ratings and half a dozen Chinese mess boys made it to the Nationalist-held Rose Island just off the Yangtze's southern bank before the evacuation was halted.

There now remained on board some seventy-six unwounded, twenty-five wounded and seventeen dead. With the evacuation over, the communists began raking *Amethyst* with machine gun bursts and rifle fire.

When firing finally ceased at around 11 a.m. twenty-two men had been killed and thirty-one wounded. *Amethyst* received more than fifty hits and was riddled with holes above and below the waterline. Those under the water were plugged with every available item from hammocks and bedding to mess tables and timber.

At 2.30 that afternoon one of the lookouts reported a ship was in sight in the distance and spirits were suddenly raised – was it *Consort* steaming fast from Nanking and coming to *Amethyst's* rescue? As the ship got closer it became clear it was indeed the Royal Navy destroyer, but she wasn't able to offer the salvation those on *Amethyst* had so fervently hoped for. Able Seaman Calcott recalled:

We heard that HMS *Consort* was at full speed coming towards us and we prepared to be towed off the riverbank. The speed at which she was travelling rocked the ship, but we were still unable to get off the mud. *Consort* turned and continued to fire broadsides but she sustained a lot of damage and casualties so had to continue her journey to Shanghai.

With the failure of *Consort*'s rescue mission all efforts were put towards freeing *Amethyst* from the mud under her own steam and Weston sent out the following message:

> *Amethyst* still aground on Rose Island. Am attempting to make good vital damage to refloat and proceed to Nanking. Sixty approx. of ship's company including four wounded are making their way to nearest town … Casualties are about seventeen dead including doctor and sick berth attendant, twenty seriously wounded including captain.

Power was restored and engines and rudder put to work to refloat the ship but the bows remained stuck fast. After Weston ordered the bows to be lightened by ditching heavy weights on the forecastle and fuel oil from the tanks, they tried again.

'That night we jettisoned as much heavy gear as possible from the forecastle and after several tries we eventually managed to release the ship from the riverbank,' said Able Seaman Calcott. 'It was decided to continue towards Nanking and we found what was considered to be a safe anchorage.'

While getting underway again had been a remarkable achievement against the odds, *Amethyst*'s problems were only just beginning. The ship's crew were now without a river pilot, their navigational chart was in ruins, the gyro-compass out of action and their magnetic compass faulty. Navigating successfully to Nanking would require a minor miracle.

By this stage *Amethyst*'s fate was making headlines around the globe and had become of grave concern to the Foreign Office, Admiralty and Prime Minister Clement Attlee. With the world watching on, her rescue was treated as a priority in Whitehall.

The 10,000-ton cruiser HMS *London* and *Amethyst*'s sister ship *Black Swan* were dispatched to run the gauntlet of communist guns and escort *Amethyst* down-river, but like *Consort* they came under such heavy fire they couldn't break through and had to turn back, with *London* suffering fifteen killed and twenty wounded.

With hopes of rescue any time soon dashed, Weston accepted the offer of help from the Chinese Nationalists in treating the wounded and allowed a doctor and two medical orderlies on board. While they didn't carry morphine, penicillin or any other drugs, they dressed wounds and made the casualties more comfortable. So hospitable was the treatment they received, Weston decided the wounded should be taken ashore and entrusted to the care of the Nationalists.

Just as these plans for the wounded were being made those on deck heard the throbbing sound of a distant aircraft's engines. It was an RAF Sunderland flying boat coming to their assistance with medical supplies. Landing 50 yards from *Amethyst*, RAF doctor Michael Fearnley disembarked and made his way to the ship. The Sunderland landing had awoken the communist guns from their slumber and batteries began firing at both aircraft and ship – raining down so many shells that the Sunderland had to get airborne again immediately to save being destroyed or joining *Amethyst* in being trapped.

Fearnley, who was shocked by the sight that awaited him, describing it as like 'a front-line dressing station', joined the Chinese doctor in administering treatment to the wounded and providing the blood plasma, morphine and penicillin that the Chinese medical team lacked. Once this work had been completed the evacuation of the wounded began. Fearnley had wanted Weston to leave the ship, fearing the wound he'd sustained could turn septic, but Weston refused and insisted he was staying put.

While Weston might not have budged at Fearnley's behest, he soon had no choice as a more senior officer – Britain's naval attaché to Nanking, Lieutenant-Commander John Kerans – arrived at the ship to take command. To reach *Amethyst*, Kerans had undertaken something of an epic journey, trekking overland by vehicle and on foot. The night before he'd missed the ship by about ten minutes when it raised anchor just before he could get on board but now he'd caught up and taken charge.

Asking Fernley for his opinion on Weston's health, Fernley replied, 'He should be got to hospital as soon as possible, sir. His life may be in danger if he stays. I can't do anything for him here.'

'I'm afraid then,' said Kerans to Weston, 'that you'll have to go. There is a landing-craft alongside. You must go at once as the railway is likely to be cut at any moment. You'll find the military attaché waiting for you at Chinkiang.'

Kerans' first grim task on taking over was to bury the seventeen corpses still on the ship. They were each weighted with two 4-inch shells and committed to the deep.

During the next few days *Amethyst* moved several times but on each occasion, as she got underway, the batteries opened fire, forcing her to anchor. The communist army was closing in and *Amethyst*'s prospects of rescue or breaking through under her own steam appeared bleaker than ever.

'The situation now seemed desperate; there seemed no hope of movement either way, and a heavy loss of life was likely if any attempt was made,' said Kerans. 'It appeared that the communists could destroy us when they chose. Morale of all on board was still good in spite of the hopelessness of our plight. I realised, however, that it might snap at any moment and that breaking point had almost been reached.'

Diplomatic wrangling behind the scenes and direct negotiations with the communists to try and seek their agreement for *Amethyst*'s safe passage now seemed the only feasible option.

On 30 April the People's Liberation Army demanded that Britain, the United States and France withdraw their armed forces from all parts of China and as negotiations opened up the Royal Navy was left dumfounded by the communists' claim that *Amethyst* had fired first and that they'd only opened fire in response to being attacked. Of course this claim was nonsense; the British knew it, the Americans knew it – the whole non-communist world knew it, but the communists insisted this was the way of it.

Mao's forces agreed to consider letting *Amethyst* out of the trap if Kerans signed a statement saying that his ship had wrongly invaded Chinese national waters and had fired at them first. The communists argued that, as they didn't recognise any treaties between the previous Chinese government and the British, it was illegal for *Amethyst* to cruise on the Yangtze. With Kerans refusing to plead guilty and admit to a charge so clearly false, negotiations stuck and a three-month impasse commenced.

With the encircling communist army holding *Amethyst* prisoner and preventing any supplies reaching the ship, oil and food supplies were running desperately low by late June. So grim had the situation become that sheets of toilet paper had to be cut in halves and later halved again.

Eventually meagre amounts of oil and food were permitted to pass through the blockade but no headway whatsoever was being made in the negotiations for securing the ship's release.

'The only way we shall ever get out of this place is to make a run for it,' said Fernley. And run for it is exactly what they did.

Getting away quickly and undetected would be paramount to the success or failure of the dash for freedom so instead of raising anchor – which would have made an almighty noise as the anchor chain rattled back on board – *Amethyst* slipped her chain at 10 p.m. on 30 July and left the anchor behind. Canvas screens had been rigged up on deck to change the ship's silhouette to trick the communist gunners into thinking they were looking at another vessel if they picked up its outline through the darkness.

Both engines burst to life and *Amethyst* crept quietly away from the anchorage where she'd spent months in captivity, following a merchant ship that by chance happened to be passing as she made her escape. Following in the wake of this other vessel would both aid navigation and make them less conspicuous. For half an hour they steamed unnoticed but then an almighty barrage of shells and small arms fire opened up.

Kerans immediately ordered full power from both engines and raised black smoke to disrupt the gunners' visibility but the bombardment continued unabated and sunk the merchant ship *Amethyst* was following.

Now she'd been detected, speed, not stealth, was the priority and the ship was pushed to 20 knots as she steamed past Rose Island. Moving out to the middle of the river, where she was less likely to be hit by shore batteries, she was soon travelling at 22 knots – ploughing through the water like never before. Reaching the old Nationalist naval base of Kiang Yin, she came under fire again from the shore and a ship, but such was her speed not a single shell touched her.

By 2.45 p.m. *Amethyst* had travelled half of the 104-mile journey to safety and as daylight approached she only had one major obstacle left to overcome – the searchlights and heavy guns of the forts of Pao Shan and Woo Sung. One direct hit on the waterline from the forts' 6-inch guns would almost certainly sink her. At 5 a.m. every man on deck could see the searchlights arching back and forth across the river right over the area through which they'd have to pass. Kerans ordered the engine room to give it 'everything you've got – damage to engines accepted'. Miraculously they managed to avoid the full beam of the lights and passed the forts without alerting their communist occupants.

Ten minutes later the breaking dawn brought with it the welcome outline of the destroyer HMS *Concord* as *Amethyst* broke out of the Yangtze and into the open sea.

The signal she sent out informing the navy of her successful escape was soon transmitted around the world: 'Have rejoined the fleet south of Woo Sung. No damage or casualties. God save the King.'

Admiral Sir Patrick Brind, Commander-in-Chief of the Royal Navy's Far East Fleet, replied:

> Welcome back to the fleet. We are all extremely proud of your most gallant and skilful escape and that endurance and fortitude displayed by everyone has been rewarded by such success. Your bearing in adversity and your daring passage tonight will be epic in the history of the navy.

Amethyst received a rapturous reception on arriving back in Hong Kong, being greeted by sirens, all manner of local vessels and an RAF flypast. She was awarded an equally joyous welcome when she got home to Devonport on 1 November 1949.

It wasn't just the humans on board *Amethyst* who'd become famous thanks to their heroism, gallantry in adversity and thrilling escape – a feline member of crew had become something of a celebrity in his own right.

Simon, a black and white tomcat, had been found wandering the dockyards of Hong Kong in March 1948 by a member of *Amethyst*'s crew. Unwell and undernourished, he was taken on board to be fed and nursed back to health. Such was his prowess as a rat-catcher that he quickly ingratiated himself with the ship's company and rapidly gained a reputation for cheekiness by leaving dead rats in sailors' beds and sleeping in the captain's hat.

When *Amethyst* first came under fire on the Yangtze, Simon was badly wounded by shrapnel and wasn't expected to last the night but, thanks to the medics cleaning his burns and removing shrapnel, he pulled through and resumed his role as a prolific rat-catcher, while his antics provided a much-needed morale boost to the crew during their time in captivity.

Following the ship's escape he was lauded for his conduct during what had been dubbed 'the Yangtze incident' and presented with the Dickin Medal (the animal version of the VC), a Blue Cross medal, an *Amethyst* campaign medal and his own naval rank of Able Seacat. Thousands of letters were written to him and he was treated as a guest of honour at every port *Amethyst* called at on her way back to Devonport.

Like all animals entering the UK, Simon was subject to quarantine regulations and spent his time at an animal centre in Surrey. While at the centre he caught a virus as a result of his war wounds and passed away on 28 November 1949.

Hundreds, including the entire *Amethyst* crew, turned out for Simon's 'funeral' in East London and they made him a headstone that read:

IN MEMORY OF "SIMON". SERVED IN HMS *AMETHYST* MAY 1948–SEPTEMBER 1949. AWARDED DICKIN MEDAL AUGUST 1949. DIED 28 NOVEMBER 1949. THROUGHOUT THE YANGTZE INCIDENT HIS BEHAVIOUR WAS OF THE HIGHEST ORDER.

Earlier that month the ship's company had been honoured with a civic reception at London's Guildhall following a service of thanksgiving at St Martin in the Fields and a march through the streets of the capital.

The *London Evening News*, under the headline 'London's cheers hail the Yangtze heroes', described the reception thus:

It was in some ways like a second Lord Mayor's Show. Thousands of people, mostly girls and young women, ran along the sides of the crowd trying to keep up with the marching column. As the line of men passed under the railway bridge at Ludgate-circus the driver of a train standing on the bridge sent the 'V' signal in Morse on his engine's whistle.

After returning to Devonport, *Amethyst* underwent a refit and returned to the Far East for two years before being paid off into reserve in 1952.

While in reserve she played herself in the 1957 film *Yangtze Incident: The Story of HMS Amethyst* that starred Richard Todd as John Kerans.

When filming was over the curtain came down on her career and she was taken to Plymouth and broken up for scrap before the movie hit the cinemas.

Amethyst seen alongside HMS *Concord* after her escape from the Yangtze River. (Michael Pocock, MaritimeQuest.com)

Amethyst on convoy duty during the Second World War.

Amethyst arriving back home to Devonport after the Yangtze Incident. (Lt Cdr K. Stewart Hett M.B.E, R.N. Retired)

Amethyst is greeted by swarms of well-wishers as she arrives back in Devonport. (Michael Pocock, MaritimeQuest.com)

Damage *Amethyst* sustained in the Yangtze Incident. (Lt. Cdr K. Stewart Hett, MBE, RN. Retired)

CHAPTER NINE

HMS *Bulwark*

HMS *Bulwark*
22,000 tons
Built by Harland & Wolff, Belfast
Served 1954–1981

She launched airstrikes in Suez, delivered commandos into Borneo and
provided flood relief in Malaysia. If versatility is valued in a warship then
Bulwark was worth her 22,000-ton weight in gold.

For more than a quarter of a century she served Britain in more roles than
just about any other Royal Navy vessel has ever managed and, no matter what
the task, the ship and her crew carried it out to the exceptional standards that
have been the Senior Service's hallmark for the past 250 years.

Bulwark was undoubtedly a jack of all trades, but a master of them too.

At the height of the Second World War in 1943 the Admiralty set out a new
construction programme identifying a need for eight light fleet carriers –
vessels that were essentially an intermediate step between the costly full-sized
fleet aircraft carriers and the small, slow and lightly armoured escort carriers
that had been prevalent in the early years of the war.

The Centaur Class of light fleet carriers was to consist of eight ships – *Hermes*,
Albion, *Arrogant*, *Bulwark*, *Centaur*, *Elephant*, *Monmouth* and *Polyphemus*.

Yet soon after the requirement for these carriers was recognised, the war
swung massively in the Allies' favour, with the Red Army following up victory
at Stalingrad by pushing the Wehrmacht back right across the Eastern Front
in the summer of 1943 and the British and Americans hitting Rommel's Afrika
Korps out of Africa and making steady progress through Italy while planning
for the main assault on occupied Europe the following year.

With thoughts now turning to Germany's inevitable defeat and the vast sums
of money that would need to be saved when peace finally came, four of the
eight Centaur Class – *Arrogant*, *Hermes*, *Monmouth* and *Polyphemus* – were
cancelled and *Elephant* renamed *Hermes*.

The contract to build *Hermes* was awarded to Vickers Armstrong of Barrow-
in-Furness and *Albion* granted to Swan Hunter on the Tyne, while *Centaur* and
Bulwark went to Belfast's Harland & Wolff.

Harland & Wolff was one of Britain's biggest and most highly regarded shipyards, yet over the years had attracted a fair amount of dubious publicity after building RMS *Titanic*, the most infamous ship in history, which sank on her maiden voyage from Southampton to New York in April 1912. The yard had also come under criticism since the late nineteenth and early twentieth century for employing an almost exclusively Protestant, Unionist workforce while allegedly shunning Roman Catholics altogether, or treating the few they did take on as second-class citizens.

Despite the politics and association with *Titanic*, Harland & Wolff continued to churn out fine ships. In the Second World War alone the yard built six aircraft carriers, two cruisers and more than 100 other naval ships as well as carrying out repairs on thousands of vessels.

Laid down just two days after VE Day, *Bulwark*'s construction featured a range of new shipbuilding techniques, with electric welding being used extensively and X-ray testing of the welding to make sure it was to the required strength and high standards.

As well as the ship itself, the geared turbine machinery to power *Bulwark* was built on site at Harland & Wolff's Queen's Island Works and it was estimated that more than 250 miles of electric cable were incorporated into the build. Throughout the course of construction various alterations were made to the design to bring it up-to-date, with the addition of spacious dining halls close to the galleys and better-equipped mess decks to improve living conditions on board.

Bulwark's launch on 22 April 1948 was a high profile event and many of the prominent figures in Ulster society – including Prime Minister of Northern Ireland Sir Basil Brooke, members of the Northern Ireland Cabinet and their wives and Sir Frederick Rebbeck, Chairman and Director of Harland & Wolff – were in attendance as she slipped into the River Lagan while an RAF flying boat flew low over the shipyard to mark the occasion.

The pace of fitting-out work was stepped up in 1950 with the outbreak of the Korean War but then slowed again when it was decided she wouldn't be needed in the Far East, so it wasn't until December 1954 that *Bulwark* was commissioned and given the motto 'Under thy wings I will trust.'

Flying trials got underway in February 1955. Unusually for a Royal Navy carrier, the first deck landing wasn't carried out by a Brit but an American when Lieutenant Commander J. W. Nance, who was on an exchange appointment to the Service Trials Unit, landed an Avenger from RN Air Station Ford. During the next three days Avenger, Sea Hawk, Skyraider, Wyvern and Gannet aircraft all tested the catapult and arrestor gear by taking off and landing with a range of different fuel and weapons loads to put all the equipment fully through its paces.

In July 1955, after diplomatic goodwill visits to Norway and Sweden, the newly commissioned carrier made television history, becoming the first Royal Navy ship to be featured in a live TV broadcast. The two-part BBC programme showed and explained to viewers how aircraft were launched and recovered and even included Sea Hawks carrying out a mock strike on the destroyer HMS *Saintes*.

Just over a year after her acclaimed appearance in the media spotlight *Bulwark* found herself heading off to war in Suez.

Opened in 1869 after ten years of construction, the Suez Canal is an artificial waterway cutting through the Egyptian desert to link the Mediterranean and Red Sea. Prior to its opening, ships travelling between Europe and Asia had to take a route around the southern cape of Africa, but the canal slashed that journey by more than 4,000 miles.

Built by France in co-operation with Egypt, the British, suspicious of growing French influence in the region, had initially been hostile to the project and declined an offer of shares in the venture. However, when the Egyptians got into financial difficulties and put their shares up for sale, the British government swooped and bought Egypt's shareholding for £4 million in the mid-1870s.

Having finally recognised the waterway's importance as a trade and military route between Britain and her colonial possessions in Asia and Australasia, the British controlled the canal for more than eighty years until it was nationalised in 1956 by the new Egyptian leader Abdel Nasser.

In a bid to regain western control of the canal following Nasser's nationalisation, Britain, France and Israel concocted a secret plan whereby Israel would invade the Egyptian-controlled Sinai Peninsula to the east of the Suez Canal and Britain and France would then call for Egypt and Israel to cease hostilities and use their refusal to lay down arms as an excuse to send in troops to recapture the canal under the pretence of keeping the warring sides apart.

So this was the situation *Bulwark* steamed into when she was dispatched from Portsmouth to the Mediterranean on 6 August 1956 with thirty Sea Hawks from 804, 810 and 895 Naval Air Squadrons embarked.

Joining her sister ship HMS *Albion* and the larger fleet carrier HMS *Eagle* in the Anglo-French naval armada – the largest to be put to sea since the Second World War – *Bulwark* began air operations against Egyptian targets.

The Sea Hawk had been introduced into Royal Navy service in March 1953 and, although it was designed as a single-seat jet fighter, it excelled in the ground-attack role it was to take on during the Suez Crisis. Between 31 October 1956 and the ceasefire on 6 November, which came about as a result of the United States's objection to the operation, Sea Hawks from *Bulwark* flew nearly 600 sorties, destroying numerous tanks, a gun battery at Port Said and three motor torpedo boats.

The operation might have ended in embarrassment for Britain and France, with America and the Soviet Union essentially forcing them to withdraw, but HMS *Bulwark*'s stock had risen from her involvement in the campaign and she and *Eagle* showed just how effective light fleet carriers could be.

In 1958, still fresh from her personal success at Suez, *Bulwark* steamed around the world, starting out in the Caribbean before going on to Halifax, Nova Scotia, and then to Gibraltar. From Gibraltar she headed for Singapore, the Royal Navy's home in the Far East, then Hong Kong and Mombasa in Kenya. While berthed at Mombasa four of her crew successfully scaled Africa's highest peak, Mount Kilimanjaro.

Towards the end of her stay in East Africa a crisis erupted in the Middle East and, instead of sailing for Madagascar as planned, she embarked troops and equipment and headed to Aden on 20 July, finally arriving back in Portsmouth on 20 November.

The Earl of Selkirk, who served as First Lord of the Admiralty between 1957 and 1959, had visited *Bulwark* during the world cruise and paid gushing compliments to the ship in a letter following his visit. He said:

> The precision and efficiency which was displayed by all those who used the flight deck, be it the pilots who took off and landed, the aircraft handlers or the catapult and mirror sight operators has no parallel elsewhere.
>
> The combination of team work which is controlled by 'Flyco' and the bridge is a supreme example of what can be achieved by human intelligence, modern equipment and a high degree of training. The result is a ship which is not only a highly efficient and potent weapon of war, capable of performing a wide variety of tasks, but a ship also able to make a very significant contribution towards maintaining the good name of the Royal Navy and Great Britain wherever you may go.
>
> I wish you all every success in the future, and I am confident that the name *Bulwark* will always stand for what is best in the Royal Navy.

One of the crew members on board during the world cruise was air mechanic Tony Atkins. Atkins joined the Royal Navy in the mid-1950s and spent much of his twelve-year naval career serving on 'The Rusty B'. He recalled:

> The world cruise of 1958 was an eye-opener for me, a young lad from Watford who thought he knew it all. Many of us were new hands and quite carefree and inexperienced. We all had our jobs to do and did them well and with good grace. I worked with the other lads from 29 Air Engineering Department mess at keeping our aircraft complement serviceable and I operated the large crane on the flight-deck, nicknamed Jumbo.

On that world cruise the crew of *Bulwark* had also witnessed the largest hydrogen bomb detonated by Britain during a nuclear test on Christmas Island in April 1958. Atkins said:

We all felt excited onboard *Bulwark* when the captain tanoyed us to tell us 'we could all see history in the making' ... We were not issued with protective gear of any kind and were not told about radioactive fallout. In retrospect I believe that the officers were unaware of the possible dangers. We were in our working clothes with bits of smoked glass or sunglasses to shield us from the promised flash. We felt a whooshing noise in our ears then a brisk breeze and a massively bright flash that seemed to go right through us. We turned and saw a massive white cloud like a tornado with a large head forming like a mushroom. There was a bright light in the cloud and a warm wind blowing past us.

When she arrived back in Portsmouth it was the end of *Bulwark*'s time as a fixed-wing carrier but far from the end of her prestigious Royal Navy career. The landing of 45 Commando Royal Marines at Port Said during the 1956 Suez crisis by an improvised helicopter force drawn from the navy, RAF and army flying from training carriers reinforced the need for a designated British helicopter assault carrier that could discharge marines and soldiers rapidly by sea and air. To meet this requirement it was decided to convert *Albion* and *Bulwark* from fixed-wing carriers to commando-carrying assault ships. With *Albion* still at sea completing a fixed-wing commission, *Bulwark* was the first of the sisters to undergo transformation.

Work on the twelve-month conversion got underway at Portsmouth Dockyard in January 1959 to remove the catapult, arrestor wires and eight 40-mm anti-aircraft guns, and provide space for landing craft and living accommodation for 600 commandos.

With the conversion, *Bulwark*'s main aircraft would become sixteen Westland Whirlwind and, later, Wessex helicopters.

The Westland Whirlwind was a British licence-built version of the US Sikorsky H-19 Chickasaw and was introduced into service with the Royal Navy in 1954. With a top speed of 110 mph the Whirlwind's primary role was to transport marines and their kit into battle but it could also carry torpedoes and depth charges and operate in an anti-ship/anti-submarine role.

When completed *Bulwark* was the first of her kind in the British armed forces and reinforced the traditionally close association between the Royal Marines and Royal Navy by giving the marines their own vessel for projecting force from the sea.

Commissioned as a commando carrier on 19 January 1960, *Bulwark* embarked men of 42 Commando and 848 Squadron's Whirlwinds before heading off to the Mediterranean for training exercise and then on to Singapore for a deployment in the Far East that included stops in the Philippines, Borneo and Hong Kong.

The training and experience gained in the Far East was put to use for real in June 1961. While sailing for hot weather trials in the Persian Gulf, *Bulwark* was diverted to Kuwait when Kuwait's much larger more powerful neighbour, Iraq, moved thousands of troops south towards the Kuwaiti border, threatening the tiny oil-rich sheikhdom with invasion. Alerted to Iraqi troop movements, the British military machine responded rapidly, sending the aircraft carrier HMS *Victorious* and *Bulwark* to the region. Whirlwind helicopters transferred the 600 men of 42 Commando ashore so they could secure an uncompleted airstrip outside Kuwait City to use against Iraqi forces should the invasion be launched. Britain's show of force had the desired effect as, although Iraq's Prime Minister refused to withdraw his country's claim to Kuwait, he gave assurances he wouldn't try to enforce it by arms.

The success of the British operation and *Bulwark*'s deployment of 42 Commando proved how effective commando carriers could be and the important role they had to play in modern warfare.

With Suez and Kuwait under her belt, *Bulwark* was becoming something of a specialist in small-scale wars in former British colonies and this trend continued in 1964 when she took part in the Borneo Confrontation – a violent three-year conflict sparked by Indonesia's opposition to the creation of Malaysia, which had been formed with the amalgamation of the former British territories of the Federation of Malaya, Singapore, North Borneo and Sarawak.

Having spent much of 1963 undergoing an extensive refit in Devonport Dockyard to service and repair equipment and improve the facilities for handling aircraft to enable her to operate the larger, more capable Wessex helicopter, which had been in service for just two years, *Bulwark* left Plymouth bound for Borneo in March 1964.

Bulwark's task was to provide the marines with the speed and mobility needed to enable them to effectively combat Indonesian infiltrators in the jungle areas of the Malaysian/Indonesian border.

Helicopters had already proved their value in jungle warfare when British forces had fought communist insurgents during the Malayan Emergency. As more and more communists were rounded up they, like the Viet Cong a decade later, retreated further into the jungle where ground vehicles and fixed-wing aircraft couldn't operate. Helicopters, however, could land in jungle clearings and were used extensively to get troops, including the SAS, into prime positions to attack the rebels.

Britain's previous experience of operating helicopters in the south-east Asian jungle proved valuable in the Borneo Confrontation. Half of 845 Squadron remained on detachment ashore, with the other half on *Bulwark*. The land-based detachment and their Wessex helicopters frequently worked alongside the SAS, Gurkhas and the Parachute Regiment to patrol a vast area of 25,000 square miles.

The difficulties of operating in such a challenging climate hit home with a series of accidents between February and April 1965 that resulted in the loss of five aircraft, three pilots, two aircrew and eleven soldiers – but *Bulwark* and her helicopters kept on fighting. Their efforts won them the Boyd Trophy – a prize presented annually for the finest feat of naval aviation.

The trophy citation, written by the Flag Officer Naval Air Command, Vice Admiral Sir Richard Smeeton, read:

> By their operations and exercises, both ashore and afloat, 845 Squadron set a standard of efficiency in Borneo unsurpassed by any other squadron deployed there and made a considerable contribution to the defence of Malaysia. They earned the praise of all. Their flying was of the highest order and the enthusiasm and professional ability of the maintenance ratings enabled the squadron to achieve an outstandingly high serviceability rate, and the distinction of meeting every demand made upon it.

On her return from the Far East in September 1965, *Bulwark* was refitted at Devonport Dockyard before returning to service in April 1966. Shortly after returning to the fleet she tested a new aircraft that would leave an indelible mark on aviation and the Royal Navy.

Since the late 1950s Hawker Aviation had been working to develop a fixed-wing aircraft that could take-off and land vertically and these experiments led to the creation of the Hawker Kestrel. The Kestrel made its first flight in March 1964 and was a major coup for Britain. With the vast Soviet air force expected to target NATO airfields in the first wave of any attack, having aircraft that didn't require conventional runways and could be hidden near the front line to operate either in an air-to-air or air-to-ground capacity from practically any piece of land would be invaluable to NATO.

As well as operating from small spaces on land, short take-off vertical landing aircraft also had the advantage of being able to take-off and land on ships without the need for large aircraft carriers with catapults and arrestor cables.

Bulwark carried out a series of trials, landing and launching a Kestrel in June 1966, with the trials going well. A slightly modified and updated version of the Kestrel was to enter service in the guise of the Harrier and Sea Harrier – aircraft that went on to play a crucial role in Britain's Falklands victory. Had *Bulwark* not successfully tested the Kestrel, the Harrier might never have entered Royal Navy service.

While on deployment yet again in the Far East during 1967, the Malaysian state of Trengganu was hit by devastating floods that claimed lives, destroyed homes and cut off remote communities from the outside world. The only way to deliver food, supplies and medical aid to some of those stranded was by air, so 845 Squadron's Wessex helicopters – now the Mk 5 version, which had two engines, giving it greater power than the one-engine Mk 1 – got to work providing vital assistance to the flood relief effort.

The next six months were spent on various exercises with the embarked 42 Commando and other regiments, interspersed with time alongside in Singapore and Hong Kong. A visit to Hong Kong in June 1967 offended communist China's sensitivities, with the Chinese Communist Party's propaganda sheet, the *People's Daily*, branding *Bulwark's* presence and the manoeuvres undertaken by her Royal Marines ashore as 'a brazen provocation at the very gate of South China'. This accusation no doubt had far more to do with it being the height of the Cold War and China's desire to see Hong Kong returned to Chinese control than it did with genuine fears about British forces in the region posing a threat.

Throughout the 1950s and 1960s, Britain's colonies had rapidly gained their independence and *Bulwark's* next role was to withdraw British forces from Aden towards the end of 1967. Aden, located at the southern end of the Red Sea, had been a British protectorate since 1839 and was an important strategic staging-post between the Mediterranean and Britain's Indian empire. In 1963 Marxist paramilitaries rose up against British rule, starting their campaign by throwing a grenade at a group of British officials at Aden airport. The uprising led to a fierce four-year fight between the rebels and British security forces that only ended when the British withdrew in late 1967. *Bulwark* embarked 45 Commando in Aden – and as the unit had been in the now former colony for seven years their equipment filled practically every available space, nook and cranny on the 22,000-ton ship.

On returning to the UK in March 1968 *Bulwark* was attached to the NATO Fleet – a NATO force aimed at defending Europe's northern flank to deter, and in the event of the Cold War turning hot, battle Soviet forces. In the event of a Soviet invasion of western and northern Europe the Royal Marines were tasked with defending Norway and as such were Arctic warfare experts. Exercise Polar Express in the summer of 1968 saw *Bulwark*, with 45 Commando on board, join a host of British and other NATO ships to land a commando force by helicopter and landing craft on the Norwegian coast to simulate NATO's response to a Soviet invasion.

The next few years consisted of regular deployments all around the world – from the Caribbean to the Far East and South Africa to Scandinavia. The biggest drama that occurred in this period was when, berthed in the Italian port of Trieste in the Adriatic, a large fire broke out and ravaged one of *Bulwark's* boiler rooms. For their efforts in protecting crew and battling the flames, Acting PO MEM Joseph Thompson of Newcastle and MEM 1 Peter Simcock from Cheshire were awarded British Empire Medals for gallantry. The blaze had taken hold after a boiler fuel hose burst, spraying hot oil and igniting the bilges and boiler. The crew of five had almost extinguished the fire when it reignited and blocked their exit route but PO Thompson remained calm and shut down machinery to stop the fire spreading, even though they were trapped by thick, choking smoke and searing heat.

When the alarm was raised, nineteen-year-old MEM Simcock burst through the flames in his fire-fighting suit to reach those trapped despite other firefighters being driven back. The heat was so intense that Simcock suffered burns through his protective gear but he battled through and everyone was rescued. Simcock's citation read:

> Simcock, only nineteen years old, spent twenty minutes almost alone in a compartment filled with heat, flames and smoke of a major fire and carried out the task he had been given ... The trapped men were recovered unharmed and it became possible to close down the compartment and fight the fire in time to stop it getting out of hand. MEM Simcock's courage, devotion to duty and coolness in the face of considerable danger were in the highest traditions of the Service.

Having tested the Kestrel back in 1966, 1973 saw *Bulwark* test the aircraft that had stemmed from the Kestrel design – the Harrier. Two RAF Harriers took off vertically from her flight deck, proving that Harriers could operate effectively at sea – a discovery that helped make up the government's mind in deciding to pursue a naval version of the aircraft which became the Sea Harrier.

By the mid-1970s *Bulwark* had been in service more than twenty years and was beginning to show her age. The size of the Royal Navy was also rapidly shrinking as successive governments slashed defence spending, so in 1976 she was put into a type of reserve known as 'preservation by operation'. This essentially entailed mothballing the ship or putting her into hibernation with a crew a couple of hundred strong remaining on board to keep her maintained to such a standard that she could be quickly reactivated if required. In 1976 the Peruvian Navy expressed an interest in buying *Bulwark* and Harrier aircraft to fly from her but the sale failed to materialise as the Ministry of Defence was unable to give a firm date for when they'd be prepared to sell her.

The paying-off of the final fleet carrier, HMS *Ark Royal,* in 1978 and the new smaller Invincible Class carriers not being ready for service until the early 1980s left *Bulwark*'s ageing sister *Hermes* as the only carrier in the fleet. To help tide the navy over until HMS *Invincible* came into service, *Bulwark* was reactivated into a fully operational state in February 1979. Instead of being a commando carrier, however, *Bulwark*'s primary task would now be to operate Wessex and Sea King helicopters in an anti-submarine role, though she did retain her ability to carry and deploy Royal Marines.

The intention had been to keep her in service for five years after reactivation but in the end she only lasted two – berthing at Portsmouth for the final time in March 1981 as the defence cuts implemented by Margaret Thatcher's new Conservative administration began to bite. For twelve months *Bulwark* lay in a corner at Portsmouth Dockyard being de-fuelled and de-stored and had parts stripped from her for use in other ships.

When the Falklands Conflict broke out in April 1982 Britain only had two active aircraft carriers – *Hermes* and *Invincible* – and such was the importance of providing the fleet with air cover in the South Atlantic that there was talk of reactivating her for use in the conflict zone, possibly as a floating barracks and heliport for British troops. However, a rapid survey of the ship discovered she was in such a poor condition that sending her south was impossible.

In April 1984, four tugs took her in tow and eased her away from her moorings for the last time to take her on a final journey to Cairnryan in south-west Scotland. Bad weather in the Irish Sea gave the 'The Rusty B' a brief stay of execution as she pulled on the lines attaching her to the tugs and delayed the journey, almost as if she was deliberately trying to delay the fate that awaited her. Entering Cairnryan, she was berthed between two forlorn looking twisted and torn frigates part way through the scrapping process that *Bulwark* was about to experience.

Bulwark at Malta in December 1956. (Michael Pocock, MaritimeQuest.com)

HMS *Bulwark* in June 1955. (Robert Edmunds)

CHAPTER TEN

HMS *Invincible*

HMS *Invincible*
22,000 tons
Built by Vickers Shipbuilding Limited, Barrow-in-Furness
Served 1980–2005

After little over eighteen months' service, *Invincible* was deemed surplus to requirements and due to be sold off – another victim of swingeing defence cuts being pushed through by a new prime minister desperate to rein in public spending.

Had a collection of remote, windswept islands in the South Atlantic not been invaded five weeks after the Australian government announced in February 1982 that it had agreed a £175 million deal to buy her, *Invincible* would have been one of the Royal Navy's shortest-serving and least celebrated warships.

Instead, by playing a key role in Britain's successful recapture of the Falkland Islands, she had greatness thrust upon her and was rewarded with cancellation of the sale and a place in the hearts of a grateful, victorious and revitalised nation.

In the early 1980s Britain was in seemingly terminal decline. Rising unemployment, an economy in turmoil, rioting in cities across the land and chaos, death and carnage on the streets of Northern Ireland saw Britain frequently dubbed the 'sick man of Europe'.

Having swept to power in 1979 on a manifesto of rolling back the frontiers of the state with a free market economy, low taxation and curbing the power and influence of trade unions to arrest Britain's decay, Margaret Thatcher was starting to look decidedly like a one-term prime minister.

Her policies seemed to be having little effect on improving the nation's fortunes and many argued Britain's situation was getting worse, not better, under Thatcher's premiership.

Then, in April 1982, her salvation came from an unlikely source. Until that month and year, few people in Britain had any idea where the Falkland Islands actually were – somewhere off the coast of Scotland was the running joke at the time.

A British overseas territory for 150 years, Argentina's ruling right-wing military junta invaded the islands the Argentines referred to as Las Malvinas and had long claimed as theirs.

In flagrant breach of the United Nations charter on the right to self-determination, and against the wishes of the 2,000 staunchly British Falkland Islanders, Argentina landed 600 troops who, after facing heroic resistance from the fifty-seven-strong Royal Marine garrison, tore down the Union Jack and hoisted the Argentine triband over the islands' capital, Port Stanley.

Argentina's dictator Leopoldo Galtieri and his commanders were convinced the British would neither have the resources nor the stomach to fight for an old colonial outpost most of its citizens had never even heard of.

As many a political friend and foe could already testify, Mrs Thatcher however was not a woman to mess with or underestimate and, having been urged by the First Sea Lord Admiral Sir Henry Leach that Britain could, and must, assemble a large naval task force to recapture the islands, she came out fighting and gave Argentina an ultimatum: withdraw completely from the Falklands or face military action as the British were coming.

Just as it did with *Invincible*, victory in the South Atlantic transformed Thatcher's fortunes and sent her approval ratings rocketing. Seen as the key architect of Britain's triumph, whose nerve, resolve and strong leadership had been crucial to the operation's success, she went from looking like a lame-duck prime minister to winning the 1983 general election by a landslide and going on to implement the policies and reforms she believed would change Britain for the better.

From being the sick man of Europe, the UK suddenly felt like a different place after the Falklands Conflict. Patriotic fervour, whipped up in no small part by the tabloid press, had swept the nation and crushing the invaders in a campaign fought 8,000 miles from home restored pride, prestige and confidence to a country that had been filled with self-doubt.

Without *Invincible* winning the war would have been highly unlikely, if not impossible, so the decision to press ahead with building her ten years earlier when Britain was supposed to have dispensed with aircraft carriers was a great stroke of luck.

'End of fixed-wing flying in the Fleet Air Arm' and 'End of navy as a global force' screamed the angry national newspaper headlines on Wednesday 23 February 1966, when Harold Wilson's Labour government cancelled the Royal Navy's planned two new large fleet aircraft carriers and announced that the four existing carriers would be phased out during the 1970s as the country looked to slash defence spending in view of its reduced role in the world.

Following the Second World War, an exhausted and bankrupt Britain could no longer afford to maintain and police a global empire so the former colonies and dominions were rapidly granted independence in the new post-war world.

The loss of empire and subsequent withdrawal from land, sea and air bases east of Suez in the mid-1960s gave the government the excuse to argue that with the Royal Navy's focus now being defence of the North Atlantic against Soviet submarines and aircraft, rather than operating in every ocean, carriers were an expensive luxury Britain no longer needed. Instead it was envisaged that RAF aircraft operating from land bases would take on the role of providing strike capability and air defence of the fleet.

For a navy with a long and proud history this was such a controversial, devastating blow that First Sea Lord Admiral Sir David Luce and Navy Minister Christopher Mayhew resigned in protest at the cancellation of the new fleet carriers.

Before the cancellation, the navy had planned to build 'through deck cruisers' that could carry helicopters and would accompany the fleet carriers on anti-submarine duties.

With anti-submarine warfare a priority for the shrinking service, even without giant new carriers to protect, plans and designs for these new ships continued and in 1967 a 12,500-ton vessel, with a cruiser design which could accommodate six Sea King helicopters, was proposed and government approval for continued planning granted.

By 1971 this 'through deck cruiser' design had been enlarged to 19,000 tons with a capacity for seventeen Sea Kings and around a dozen new Hawker Siddley Harrier V-STOL aircraft that could take-off and land vertically. With the development of this revolutionary plane, the navy saw a way of maintaining a fixed-wing capability through the back door as the beauty of the Harrier was that being able to land vertically meant it could operate from far smaller carriers than conventional aircraft.

With the trials of the new Harrier having been successfully carried out on Britain's last two remaining fixed-wing carriers, *Eagle* and *Ark Royal*, in 1972 the government awarded a contract to Barrow-in-Furness-based Vickers Shipbuilding Limited to finalise designs and begin work on the new ship, which would be the largest built for the Royal Navy for more than a quarter of a century.

Her keel was laid in July 1973 and she was launched by Her Majesty the Queen with the breaking of a home-made bottle of wine on her bow on 3 May 1977.

In the commemorative booklet for the launch her builders said:

Vickers are proud that one more prototype – 'the difficult ships to build' – takes to the water from their Barrow shipyard, giving to the Royal Navy a ship which is in every way designed and equipped for naval operations in the 1980s, a ship that is not only another first of class, but the best which British technology can give the Senior Service.

While *Invincible* was under construction in Barrow a naval version of the Harrier – the Sea Harrier – had been developed for the Fleet Air Arm to equip *Invincible* and the ageing light fleet carrier turned commando helicopter carrier

Hermes, which with the arrival of the Sea Harrier became a fixed-wing carrier yet again.

Although, in the 1970s, the Harriers weren't seen as a replacement for the fleet's dependence on the RAF for its main air cover and strike capability, they did at least ensure the Royal Navy remained one of only a handful of navies capable of deploying fixed-wing aircraft. The plan was for the Sea Harrier to provide limited air defence and anti-ship capability.

Following *Invincible's* launch she was moved to Vickers' Buccleuch Dock, where 3,500 workers swarmed all over her for the ship's fitting-out. Her completion involved installing a million components served by a thousand kilometres of cables and a hundred kilometres of pipes. Unlike earlier British aircraft carriers equipped with boilers and steam turbines, *Invincible* was powered by four Rolls-Royce gas turbines generating 97,000 hp that could drive her through the water at a top speed of 28 knots.

As well as Sea Harriers and Sea King helicopters, she was armed with the new Sea Dart surface-to-air missile, Goalkeeper close-in weapon system and Oerlikon cannons and had a 'ski-ramp' fitted at the forward end of her flight deck so the Harriers could be launched on an upward trajectory, enabling them to take-off with heavier fuel and weapon loads. At £215 million she was then the most expensive ship ever built for the Royal Navy.

Invincible left Barrow on 26 March 1979 for dry-docking at Greenock and her initial sea trials, which included the first helicopter deck landing, off Scotland's west coast. She then returned to Vickers for final tests and fitting-out completion before heading to her new home port of Portsmouth, where she was commissioned, with the Queen in attendance, in July 1980.

Later that year the Sea Harriers of 800 Squadron, together with Sea King helicopters, joined *Invincible* for an intense six weeks of flying trials where they carried out launchings and landings in all weathers, day and night, to test the aircraft and ship to their limits.

In early 1981 she carried out joint exercises with the French fleet, including the aircraft carrier *Clemenceau*, in the Mediterranean and then went on to conduct operational sea training off Portland with the embarked 801 Squadron's Sea Harriers. With this training completed *Invincible* then returned to Portsmouth as a fully operational unit of the fleet.

At this point she should have had a twenty-five to thirty-year Royal Navy career stretching out in front of her, but instead Thatcher, Defence Secretary John Nott and their Whitehall bean-counters were already planning to dispense with her as part of navy cuts that went even deeper than those implemented by the Labour government in the mid-1960s. As we've already seen, Galtieri saved her just in the nick of time.

When news came through that British sovereign territory had been invaded for the first time since the Japanese marched through Britain's far-eastern possessions in 1941 and 1942, *Invincible* was berthed in Portsmouth with much of her crew dispersed all across the UK and even further afield on leave.

In the early hours of Friday 2 April an urgent signal was received on board ordering the ship be brought to four hours' notice for sea, so her company had to be recalled by every available means – via BBC local radio stations, telephone calls to their homes or contacting the addresses left on their leave cards.

Given the distances involved between the UK and the South Atlantic, the upcoming operation was the biggest logistical challenge British forces had faced since 1945 – and this time they'd be going it alone without the massive contribution and support of two super-power allies that they'd had in the Second World War.

The location of the Falklands, well away from any potential friendly land bases, meant the campaign to oust the Argentinians would fall primarily to the Royal Navy. The navy was the only service able to transport men, munitions and equipment to the islands and the only one capable of winning the sea and air supremacy needed if the British landings were to have any chance of success.

As well as the aircraft carriers, destroyers and frigates that could take the battle to Argentina, a vast armada of transports and cargo ships was needed for the operation.

To supplement the Royal Fleet Auxiliary's tankers, landing ships and supply vessels, the Ministry of Defence requisitioned more than forty Merchant Navy ships in a move called STUFT (Ships Taken Up From Trade).

The STUFT vessels included the liners *Canberra* and *QE2* that were used to transport 3 Commando Brigade and 5 Infantry Brigade respectively, and the North Sea ferry *Norland* which carried 2 Para south. Roll-on-roll-off ferries *Elk*, *Baltic Ferry* and *Europic Ferry* became ammunition and vehicle carriers while the container ships *Atlantic Conveyor* and *Atlantic Causeway* transported helicopters, equipment for the troops ashore and additional Harriers.

Requisitioning the merchant ships was the easy part – they then had to be converted for use in a war zone and fitted with helicopter landing pads, military communication systems and anti-aircraft guns.

Dockyard workers, many of whom ironically had received their redundancy notices just days earlier as part of the defence cuts, toiled around the clock to get *Invincible* and the Task Force's vast array of other ships ready for the 8,000-mile voyage. Not since the Allies prepared for the D-Day landings had the south of England been such a hive of military activity.

Invincible's Captain J. J. Black recalled how preparing his ship for a potential war was quite literally a case of all hands on deck:

The scene on board resembled an ants' nest, brisk but purposeful, with all sorts of people, civilian and military, hastening to their next assignment. This tensely focused atmosphere was to prevail until ultimately we sailed on Monday morning.

Every available body, be he a loaned sailor from another ship, unoccupied aircrew, the dentist or the padre – was dressed in overalls and helping to transport the tons of stores and ammunition which came over the side by every gangway, chute or crain.

Winning air superiority in the seas around the Falklands would be a priority so the Task Force was centred round the Royal Navy's only two remaining aircraft carriers – *Hermes* and *Invincible* – and the, as yet untested in combat, Sea Harriers they carried.

Six destroyers – five of them the new Sea Dart-equipped Type-42s – and seventeen frigates headed for the South Atlantic alongside the carriers and two landing platform docks, *Fearless* and *Intrepid*. A day before the bulk of the Task Force set sail, the nuclear submarine HMS *Conqueror* put to sea. Unseen and unheard, she made her way silently south.

As well as the Sea Harriers of 801 Squadron, 820 Squadron and their Sea King helicopters were embarked on *Invincible* for the campaign. One of the Sea King pilots would have been a most prized prisoner had he fallen into Argentine hands. Despite the government's apprehension about the propaganda prize it would hand the enemy if he was killed or captured, twenty-year-old Prince Andrew stayed with his squadron.

The scene at Portsmouth as the Task Force raised steam turned the clock back to the glory days of empire – Union flag-wielding crowds lined vantage points around the harbour to wave a big British armada on its way to claim territory for the Crown. *Invincible*'s Skipper J. J. Black recalled:

> Huge numbers of well-wishers could be seen covering the length of Southsea front ... The scene brought a lump to the throat. These thousands of ordinary people had travelled many miles to wish us God's Speed. I felt a heavy weight of responsibility. The people of Portsmouth had acted similarly as Nelson boarded his barge in Old Portsmouth, nearly 200 years before. Implicitly they were putting their faith in us, while we had little idea at that moment of the task ahead or our ability to succeed.

Even after the Task Force had set sail few thought all-out war likely as the United Nations and United States frantically sought to avoid bloodshed by bringing Britain and Argentina to the negotiating table.

US Secretary of State Al Haig engaged in a concerted shuttle diplomacy effort, travelling repeatedly between Buenos Aires, London and Washington in hope of finding a peaceful resolution to the crisis. With neither Thatcher nor Galtieri willing to back down, peace was a forlorn hope.

Invincible's voyage very nearly ended in shame and disaster just a few miles out of Portsmouth and put a stop to the whole campaign, given she was one of only two aircraft carriers and would be vital if the operation was to succeed.

J. J. Black went on:

Commander Mike Kendrick, the Engineering Officer, drew me to one side; 'Sir, we have heard suspicious knocking from the starboard coupling in the gearbox.' My heart sank; were we to let down our nation in its hour of need? There was no one who could take our place and we had previous experience of a knock in a gearbox coupling. It meant two weeks alongside with the ship absolutely still, while dockyard engineers opened a gearbox the size of a small house and exchanged a three-ton coupling. For a start a replacement, if indeed one was available, would be in the works of Rolls-Royce in Derby. Had so many struggled so willingly over the past days in order to flounder at the eleventh hour? Would we be the straw that broke the camel's back of this courageous deployment? How would Maggie Thatcher react? How the navy would be slated by the press, those instant experts! It was all an unthinkable scenario.

With so much resting on *Invincible*, the type of stops that can only be pulled out in war, or a possible build up to war, were pulled. Within twenty-four hours a replacement coupling had been transported from Derby to Culdrose in Cornwall, flown out to *Invincible* in the Western Approaches by an RAF Chinook helicopter and a potential disaster had been averted.

About halfway between Britain and the Falklands, 1,000 miles off the west coast of Africa in the middle of the Atlantic, lies an isolated volcanic outcrop. Ascension Island is part of the British Overseas Territory of Saint Helena, Ascension and Tristan da Cunha and had long been a safe haven and coaling station for the Royal Navy. In 1982 the island became the staging post for Operation Corporate – the codename for Britain's recapture of the Falkland Islands.

The majority of the Task Force stopped off at Ascension for resupply, transferring of fuel and ammunition and time ashore for the Royal Marines and soldiers to carry out live weapons training.

Although the island is British, in 1982 its large airfield was used primarily by the United States Air Force and NASA as a tracking station for American space operations. When the British military machine took it over on its way to the South Atlantic, the airfield temporarily became the busiest airport in the world as Royal Navy helicopters and Harriers and RAF tankers, transports and Vulcan bombers moved in and out around the clock.

Invincible arrived at Ascension on 16 April, ten days after leaving Portsmouth, and joined the frigates *Alacrity*, *Broadsword*, *Glamorgan* and *Yarmouth*, fellow aircraft carrier *Hermes* and the landing ship RFA *Sir Tristram* at anchor off the island's south-west coast.

For the two days she was at anchor she embarked more stores and ammunition and practiced intercepting and repulsing mock air attacks launched by HMS *Hermes*.

Leaving Ascension behind and steaming southwards towards a South Atlantic winter, the weather got cooler, the waves rougher and conditions more unpredictable for *Invincible* and the other warships and merchantmen of the Task Force.

On 1 May she entered the Total Exclusion Zone (TEZ) – a 200-mile area in and around the Falklands the British government had imposed – within which any Argentinian forces would be considered hostile and liable to be attacked.

Now it was clear Argentina would not be going without a fight and the world watched in fascination.

While the professional British forces were undoubtedly better trained and more highly skilled than the Argentinian conscripts facing them, they were fighting 8,000 miles from home in a hostile climate with far fewer aircraft and were heavily outnumbered with the amount of troops deployed in theatre. The outcome was far from certain. *Invincible's* Captain J. J. Black continues:

> We crossed the 200 mile exclusion zone and were then technically at war at 09.00 hours … As I shaved that morning, I was contemplating the day before us and wondering whether we would have any aircraft left that evening. The Sea Harrier was of a novel design which had not been tested in battle. Our opponents flew Mirage III's which had been in action with the Israelis, and no doubt other nations besides. Surely if the Israelis had chosen to buy this French-designed aircraft, it must be a capable machine? I waited the outcome of the first few encounters with baited breath.

On the day *Invincible* entered the TEZ an RAF Vulcan bomber carried out what was then the longest bombing mission ever undertaken by travelling fifteen hours, with numerous air-to-air refuellings from Victor tankers, from Ascension Island to bomb Port Stanley airfield and make it unusable to Argentinian aircraft.

Sea Harriers from *Hermes* followed up the Vulcan mission by launching a cluster bomb attack on the airfield that was made famous by the BBC journalist Brian Hanrahan who, embarked on *Hermes*, delivered the immortal line 'I'm not allowed to say how many planes joined the raid but I counted them out and counted them all back in' as an inspired way of getting around the reporting restrictions to let viewers back home know no Harriers had been lost.

Although all the Harriers returned safely from the Port Stanley mission, many echoed J. J. Black's concerns about the aircraft's capabilities, including 801 Squadron's Commanding Officer Lieutenant-Commander Nigel 'Sharkey' Ward, who recalled:

> With our little radar in the Sea Harrier we couldn't see small, fast targets down low over a rough sea … This was a real problem and meant that the Étendards coming in, in the right conditions, could possibly sneak past us and confront the missile systems, which was a real worry to us.

While the Argentinian Mirage IIIs and Super Étendards were far quicker than the untried and untested Sea Harriers, it soon became clear that the Harriers' manoeuvrability and the American Sidewinder missiles they were armed with made them more than a match for Argentina's air force.

On 1 May, Flight-Lieutenant P. C. Barton of 801 Squadron claimed the first British air-to-air kill of the war, shooting down a Mirage. Later in the day, with three Canberra bombers approaching *Invincible*, a pair of Sea Harriers destroyed two of the enemy aircraft.

On 2 May the war suddenly escalated into bloody conflict when *Conqueror* torpedoed and sank the Argentine cruiser *General Belgrano* with the loss of more than 300 men.

Conqueror's action had the desired effect, sending the rest of the Argentinian navy – including the aircraft carrier *Veinticinco de Mayo* which, like *Belgrano*, posed a serious risk to British forces – scurrying back to port where they stayed for the rest of the campaign.

While the Argentine navy didn't show its face again for fear of meeting British nuclear submarines, the air force launched a sustained bombardment against the Task Force.

The Sea Harrier pilots flew round the clock to repulse wave after wave of attacks and, although heavily outnumbered, time and again got the better of their Argentinian foes.

Such was 801 Squadron's success rate at bringing down the attackers, it was said the Argentine air crews referred to the dark-grey painted Sea Harriers as 'La Muerta Negra' or 'The Black Death'.

Yet, despite the miracles performed by the Fleet Air Arm, the enemy still frequently got through. HMS *Sheffield* – the first Royal Navy warship lost to enemy action since the Second World War – fell victim to an Exocet missile while bombs claimed *Coventry*, *Ardent* and *Antelope*.

The container ship *Atlantic Conveyor* was also sent to the bottom by an Exocet with her cargo of Chinook helicopters. Fortunately the additional Harriers she was carrying – RAF GR3s for HMS *Hermes* and four Sea Harriers for *Invincible* – had been flown off before the missile hit.

Ardent and *Antelope* sank in San Carlos Water, the site chosen as the landing area for British ground forces, which, once identified by the Argentine air force, became known as 'Bomb Alley' for the volume and ferocity of attacks launched on British ships operating in the inlet.

So vital were *Hermes* and *Invincible* to the whole British operation that the Task Force's Naval Commander, Admiral Sandy Woodward, kept them out to sea well away from the landing ground although the Harriers were constantly vectored towards San Carlos for aerial battles.

By now, the Royal Navy pilots had honed their skills to such an extent that flying combat missions was almost becoming second nature. Lieutenant

Commander David Braithwaite, flying with 809 Squadron from *Invincible*, recalled:

> You tended to switch off the bad things because you couldn't let thinking about them affect your flying … One was conscious only of being cold, and of being tired. Initially everybody was a bit frightened, but even that wears off. That's when you've got to be careful.

While the Royal Navy suffered badly, although in truth it could have been far worse as many Argentinian bombs failed to detonate after hitting their targets, the losses the Harriers and the fleet's anti-aircraft missiles inflicted on the Argentine air force were devastating to Argentina's military capabilities and morale.

Once British forces landed on the islands, Argentinian resistance began to falter as it was soon clear the South American conscripts were no match for the supremely well-trained and professional Royal Marines and Paratroopers.

On 14 June 1982 Argentine forces in the Falklands surrendered and the Union flag once again flew over Port Stanley's Government House.

During the course of the war, the Sea Harriers shot down twenty-three Argentinian aircraft without sustaining a single loss in air-to-air combat. While this was undoubtedly down to the skill of the pilots and the quality of the aircraft they flew, Argentinian tactics were seriously open to question, as 'Sharkey' Ward recognised:

> I think their command realised that as they were quite a long way from the amphibious landing point – 300-odd miles – they wanted to maintain their assets and keep them alive so that they could deliver ordnance against the ground forces rather than play against Sea Harrier. I think it was entirely the wrong tactic and I'm quite certain the Argentine pilots would have much preferred to have taken on the Sea Harrier, outnumber it, and hopefully for them shoot a few of us down.

Argentina's other serious error was to concentrate so much of their resources attacking Royal Navy warships in and around San Carlos, rather than targeting the landing vessels and merchantmen whose losses Britain couldn't have sustained in any great numbers. Part of the job of a warship is to draw enemy fire and protect the transports around them, so in that respect *Ardent*, *Antelope*, *Coventry* and *Sheffield* performed their duties to perfection.

At the end of the war much of the victorious British fleet soon steamed north back to the UK, but *Invincible*'s duties were far from over. Tasked with providing air defence for the Falklands until relieved by her new sister ship *Illustrious*, it wasn't until late August that she could head home.

Captain Black was full of praise for every single member of crew, but knew it was his Harriers that were key to Britain's triumph:

While everyone on board played a significant role, not least 820 Squadron with their nine Sea Kings – they flew the equivalent of once around the world in often appalling weather without the loss of an aircraft – it must be 801 Squadron, the Sea Harriers, who take the ship's honours for the campaign. This small body of men, under Lieutenant Commander 'Sharkey' Ward, faced the enemy at close range. They had been reinforced by some aircraft of 899 Squadron. They shot down nine aircraft, of which 'Sharkey' shot down three, including a Hercules transport.

During the course of the war 801 Squadron flew 599 combat missions for a total of 786 hours, dropped fifty-six bombs and fired twelve Sidewinder missiles.

On *Invincible*'s rapturous return to Portsmouth, Rear-Admiral Derek Reffell summed up her achievements perfectly when he signalled: 'No other ship has contributed so much to the success of the Task Force Group operations.'

After the massive cuts looming before the Falklands, the mid-1980s proved to be something of, an albeit brief, boom time for the Royal Navy. Ships lost in the South Atlantic were replaced and the Invincible Class carriers became the focal point of the fleet.

Invincible's sale to Australia was cancelled and she was joined in service by *Illustrious* in 1982 and *Ark Royal* in 1985.

A decade of peacetime exercises and routine deployments followed her Falklands heroics, but in 1993 *Invincible* found herself in a conflict zone again – this time leading a large Royal Navy deployment to the Adriatic to provide assistance to United Nations troops ashore and police the no-fly zone over the former Yugoslavia as the Balkans was torn apart by civil war.

In 1993 the Royal Navy received an improved version of the Sea Harrier – the Sea Harrier FA2 – equipped with the more advanced Blue Vixen radar and the new AIM-120 AMRAAM air-to-air missiles – and *Invincible* returned to the Adriatic with the upgraded aircraft on board in 1994 and again in 1995.

With Bosnian Serb forces having lain siege to the city of Sarajevo, 800 Squadron's Harriers took part in the NATO air-strikes against Serb-held positions during September 1995, undertaking twenty-four bombing raids and forty-two combat patrols in the ten-day campaign.

As part of Operation Bolton, *Invincible* was deployed to the Persian Gulf in early 1998 and prepped to conduct offensive actions against targets in Iraq as a result of Saddam Hussain's failure to allow UN weapons inspectors access to suspected weapons sites. With the clock counting down to military action, a diplomatic solution was agreed at the last minute and the crisis averted.

Her final war zone appearance came in 1999 when she took part in NATO's campaign against the Federal Republic of Yugoslavia, with her Sea Harriers patrolling Yugoslavian air space to deter attacks on NATO forces.

To provide extra flight deck space *Invincible*'s Sea Dart missile system was removed before the turn of the new millennium and she then underwent a major refit between 2001 and 2003 that could have kept her in active service until well into the second decade of the twenty-first century.

Yet in August 2005, just twenty months after the refit that could've prolonged her life for another ten years, she was decommissioned.

The Royal Navy's Sea Harriers were retired in 2006 and the scrapping of the RAF's Harrier GR7s in the brutal defence cuts of 2010 brought an end to fixed-wing flying in the navy until the Queen Elizabeth Class carriers are finally brought into service, many years behind schedule.

Invincible's tale may have ended in a Turkish breaker's yard, but for what she did to revitalise a nation in the Falklands, this ship's story is one that will keep on being told.

HMS *Invincible*. (Michael Pocock, MaritimeQuest.com)

Select Bibliography

Adkins, Roy and Lesley, *Jack Tar: Life in Nelson's Navy* (2008)

Arthur, Max, *Lost Voices of The Royal Navy* (1996)

Ballantyne, Iain, *H.M.S. Rodney: Warships of the Royal Navy* (2008)

Ballantyne, Iain, *Warspite: Warships Of The Royal Navy* (2010)

Black, Admiral Sir Jeremy, *There and Back, The Memoirs of Admiral Sir Jeremy Black* (2005)

Blake, Nicholas and Richard Lawrence, *The Illustrated Companion To Nelson's Navy* (1999)

Brown, D. K. and George Moore, *Rebuilding The Royal Navy: Warship Design Since 1945* (2003)

Burns, Ken and Mike Critchley, *HMS Bulwark 1948–1984* (1986)

Burns, Ian M., *The RNAS and the Birth of the Aircraft Carrier 1914–1918* (2014)

Coleman, Terry, *Nelson: The Man and The Legend* (2001)

Darling, Kev, *Fleet Air Arm Carrier War: The History of British Naval Aviation* (2009)

Davies, Wyn and Geoff Dennison, *HMS Warrior: Ironclad Frigate 1860* (2011)

de L. W., L., *Souvenir of H.M.S. Victory* (1891)

Lambert, Andrew, *Admirals* (2008)

Lavery, Brian, *Life Series: Life In Nelson's Navy* (2007)

Lowry, Thomas P. and John W. G. Wellham, *The Attack on Taranto: Blueprint for Pearl Harbour* (2000)

Massie, Robert K., *Dreadnought: Britain, Germany and the coming of the Great War* (2013)

McCart, Neil, *Harrier Carriers HMS Invincible Vol 1* (2004)

Philips, C. E. Lucas, *Escape of the Amethyst* (1957)

Redford, Duncan and Philip D. Grove, *The Royal Navy: A History Since 1900* (2014)

Ridley-Kitts, Daniel G., *The Grand Fleet 1914–19: The Royal Navy in the First World War* (2013)

Robbins, Guy, *The Aircraft Carrier Story 1908–1945* (2001)

Schofield, B. B.: *Stringbags In Action: Taranto 1940 & Bismarck 1914* (2010)

Turner, Barry, *Suez 1956* (2006)

Wragg, David, *Stringbag: The Fairey Swordfish at War* (2004)

Websites

www.hmshood.com
www.hms-victory.com
www.hmswarrior.org
www.maritimequest.com
www.royalnavy.mod.uk

About the author

John Ballard is a journalist, writer and former press officer with a keen interest in maritime history. While working as a reporter for the *Wiltshire Times* newspaper he regularly wrote news and feature articles about the army and is the author of *Three Lions Roaring*, a book telling the story of ten classic England football team matches between 1938 and 1998.

Acknowledgements

10 Greatest Ships of the Royal Navy would never have come down the slipway, so to speak, without numerous people providing assistance and dedicating their time to help me build the book.

First of all I'd like to thank Amberley Publishing for taking on the project I pitched last autumn and particular thanks are due to Amberley's Connor Stait for editing the manuscript.

As with any book, a wide range of source material was needed to build up a picture and provide the details and background to each ship, so my friend Colin Phipps' collection of naval publications proved invaluable and I'm most grateful to Colin for the loan of his various titles.

Sourcing photos, as well as writing the text, formed a sizeable part of creating 10 Greatest Ships of the Royal Navy. I'm thankful to Peter Swarbrick and Michael Pocock for providing many of the images contained in these pages and to Elizabeth Ballard for scanning photographs for me.

Last, but by no means least, I'd like to thank my wife Mary, son Michael and daughter Katie for putting up with me disappearing into a world of sea battles, dreadnoughts and aircraft carriers while I wrote the book. I think they probably now know just as much about Trafalgar, Jutland and the Falklands as I do.

John Ballard
Cumbria, June 2015